AF595125

Image Simulation in Remote Sensing

Image Simulation in Remote Sensing

Editor

Yang Dam Eo

MDPI • Basel • Beijing • Wuhan • Barcelona • Belgrade • Manchester • Tokyo • Cluj • Tianjin

Editor
Yang Dam Eo
Konkuk University
Korea

Editorial Office
MDPI
St. Alban-Anlage 66
4052 Basel, Switzerland

This is a reprint of articles from the Special Issue published online in the open access journal *Applied Sciences* (ISSN 2076-3417) (available at: https://www.mdpi.com/journal/applsci/special_issues/Image_Simulation_Remote_Sensing).

For citation purposes, cite each article independently as indicated on the article page online and as indicated below:

LastName, A.A.; LastName, B.B.; LastName, C.C. Article Title. *Journal Name* **Year**, *Volume Number*, Page Range.

ISBN 978-3-0365-3579-1 (Hbk)
ISBN 978-3-0365-3580-7 (PDF)

Contents

About the Editor

Yang Dam Eo (Professor) received his MS and PhD at the Department of Urban Engineering of Seoul National University in 1991 and 1999, respectively. From 2003 to 2008, he worked in the Agency for Defense Development (ADD) as a senior researcher. Since 2008, he has served as a professor in the Department of Advanced Technology Fusion, Konkuk University, Seoul, Rep. of Korea. He is interested in image simulation and geospatial information extraction from satellite imagery, especially in inaccessible areas.

Editorial

Special Issue on Image Simulation in Remote Sensing

Yang Dam Eo

Department of Civil and Environmental Engineering, Konkuk University, Seoul 05029, Korea; eoandrew@konkuk.ac.kr; Tel.: +82-2-450-3078

1. Introduction

Recently, various remote sensing sensors have been used and their performance has developed rapidly [1]. Therefore, the range of remote sensing image users is expanding, and user requirements are also being advanced. In order to meet user needs, research is being actively conducted to simulate and generate remote sensing images that are limitedly acquired by various weather, environmental, and satellite operating conditions. In this issue, we deal with the research results regarding the generation of more diverse images for various environments, climates, and weather conditions, and we use them to increase the number of learning images, simulate military operations, and simulate seasonal images.

2. Image Simulation in Remote Sensing

Dae Kyo Seo and Yang Dam Eo [2] fused a panchromatic image with a SAR image to improve object recognition. By learning each class independently, the improved results were compared to existing methods. This method was designed to provide a geospatial information base without a loss of information while considering differences in the image mechanism of the two images.

Han Sae Kim and coauthors [3], in their paper, investigated kinematic in situ self-calibration to frequently re-calibrate a backpack-based MBL (Multi-Beam LiDAR) system using on-site data for handling unstable measurements of a sensor. Frequent in situ calibration prior to MBL data acquisition is an essential step in order to meet accuracy-level requirements and to implement these scanners for precise mobile applications. A simulator program was first utilized to generate simulation datasets with various observation settings, network configurations, test sites, and targets. Afterwards, self-calibration was carried out using the simulation datasets.

The high operational cost of aerial images makes it difficult to acquire periodic observations of a region of interest. Satellite imagery is an alternative for this problem and, in their article, Suhong Yoo and coworkers [4] propose a context-based approach to simulate the 10 m resolution of Sentinel-2 imagery to produce 2.5 and 5.0 m prediction images using an aerial orthoimage. This can be considered as an alternative to providing high-resolution images in a cost-effective way in the field of remote sensing

A rotational shearing interferometer has been proposed for the direct detection of extra-solar planets. This technique consists of the non-total cancellation of star radiation in order to improve signal magnitude. Beethoven Bravo-Medina and coauthors [5] propose a novel method to enhance signal magnitude by means of a star–planet interference, as well as the use of interferograms that are computationally simulated to confirm the viability of the technique.

Despite advances in SAR image processing, existing detection technologies still have limitations in boosting detection performance because of their inherently noisy characteristics. Sujin Shin and collaborators [6], in their contribution, propose a novel object detection framework that combines an unsupervised denoising network and a traditional detection network to leverage a strategy for fusing region proposals extracted from both raw SAR images and synthetically denoised SAR images.

Citation: Eo, Y.D. Special Issue on Image Simulation in Remote Sensing. *Appl. Sci.* **2021**, *11*, 8346. https://doi.org/10.3390/app11188346

Received: 3 September 2021
Accepted: 7 September 2021
Published: 9 September 2021

Changno Lee and Jaehong Oh [7], in their paper, propose sensor level mosaicking to generate a seamless image product with geometric accuracy to meet mapping requirements. The proposed method successfully identifies and removes irregular image discrepancies between adjacent data.

Acknowledgments: We would like to thank all the authors who contributed to the valuable research results and the reviewers who provided professional advice to ensure that a high-quality paper was published. Additionally, we place on record our gratitude to the editorial team of *Applied Sciences*.

Conflicts of Interest: The authors declare no conflict of interest.

References

1. Seo, D.K.; Eo, Y.D. Multilayer Perceptron-Based Phenological and Radiometric Normalization for High-Resolution Satellite Imagery. *Appl. Sci.* **2019**, *9*, 4543. [CrossRef]
2. Seo, D.K.; Eo, Y.D. A Learning-Based Image Fusion for High-Resolution SAR and Panchromatic Imagery. *Appl. Sci.* **2020**, *10*, 3298. [CrossRef]
3. Kim, H.S.; Kim, Y.; Kim, C.; Choi, K.H. Kinematic In Situ Self-Calibration of a Backpack-Based Multi-Beam LiDAR System. *Appl. Sci.* **2021**, *11*, 945. [CrossRef]
4. Yoo, S.; Lee, J.; Bae, J.; Jang, H.; Sohn, H.-G. Automatic Generation of Aerial Orthoimages Using Sentinel-2 Satellite Imagery with a Context-Based Deep Learning Approach. *Appl. Sci.* **2021**, *11*, 1089. [CrossRef]
5. Bravo-Medina, B.; Strojnik, M.; Mora-Nuñez, A.; Santiago-Hernández, H. Rotational-Shearing-Interferometer Response for a Star-Planet System without Star Cancellation. *Appl. Sci.* **2021**, *11*, 3322. [CrossRef]
6. Shin, S.; Kim, Y.; Hwang, I.; Kim, S. Coupling Denoising to Detection for SAR Imagery. *Appl. Sci.* **2021**, *11*, 5569. [CrossRef]
7. Lee, C.; Oh, J. Sensor-Level Mosaic of Multistrip KOMPSAT-3 Level 1R Products. *Appl. Sci.* **2021**, *11*, 6796. [CrossRef]

applied sciences

Article

Sensor-Level Mosaic of Multistrip KOMPSAT-3 Level 1R Products

Changno Lee [1] and Jaehong Oh [2,*]

[1] Department of Civil Engineering, Seoul National University of Science and Technology, Seoul 01718, Korea; changno@seoultech.ac.kr

[2] Department of Civil Engineering, Interdisciplinary Major of Ocean Renewable Energy Engineering, Korea Maritime and Ocean University, Busan 49112, Korea

* Correspondence: jhoh@kmou.ac.kr; Tel.: +82-51-410-4462

Featured Application: The proposed method can generate a mosaic image at the product level that is corrected only for radiometric and sensor distortions.

Abstract: High-resolution satellite images such as KOMPSAT-3 data provide detailed geospatial information over interest areas that are evenly located in an inaccessible area. The high-resolution satellite cameras are designed with a long focal length and a narrow field of view to increase spatial resolution. Thus, images show relatively narrow swath widths (10–15 km) compared to dozens or hundreds of kilometers in mid-/low-resolution satellite data. Therefore, users often face obstacles to orthorectify and mosaic a bundle of delivered images to create a complete image map. With a single mosaicked image at the sensor level delivered only with radiometric correction, users can process and manage simplified data more efficiently. Thus, we propose sensor-level mosaicking to generate a seamless image product with geometric accuracy to meet mapping requirements. Among adjacent image data with some overlaps, one image is the reference, whereas the others are projected using the sensor model information with shuttle radar topography mission. In the overlapped area, the geometric discrepancy between the data is modeled in spline along the image line based on image matching with outlier removals. The new sensor model information for the mosaicked image is generated by extending that of the reference image. Three strips of KOMPSAT-3 data were tested for the experiment. The data showed that irregular image discrepancies between the adjacent data were observed along the image line. This indicated that the proposed method successfully identified and removed these discrepancies. Additionally, sensor modeling information of the resulted mosaic could be improved by using the averaging effects of input data.

Keywords: KOMPSAT-3A; strip; sensor modeling; RPCs; mosaic; matching; discrepancy

Citation: Lee, C.; Oh, J. Sensor-Level Mosaic of Multistrip KOMPSAT-3 Level 1R Products. *Appl. Sci.* **2021**, *11*, 6796. https://doi.org/10.3390/app11156796

Academic Editor: Yang Dam Eo

Received: 22 June 2021
Accepted: 22 July 2021
Published: 23 July 2021

1. Introduction

High-resolution satellite images provide detailed geospatial information with a high geospatial resolution up to 30~80 cm over the area of interest, even located in inaccessible areas. There are many operating satellites such as Ziyuan-3 (2.1 m), KOMPSAT-2 (1 m), Gaofen-2 (0.8 m), TripleSat (0.8 m), EROS B (0.7 m), KOMPSAT-3 (0.7 m), Pléiades 1A/1B (0.7 m), SuperView 1–4 (0.5 m), GeoEye-1 (0.46 m), WorldView-1/2 (0.46 m) and WorldView 3 (0.31 m), etc. [1]. The satellites operate at low altitudes, such as 500,700 km, to achieve a high geospatial resolution of the data. In addition, the satellite cameras are specially designed by increasing the focal length up to around 10 m using a few aspherical mirrors. For example, WorldView-2, Pleiades-HR, and KOMPSAT-3 have focal lengths of 13.311, 12.905, and 8.562 m, respectively.

As a trade-off for the low altitude and long focal lengths, the high-resolution satellite data show a relatively narrow field of view compared to the mid- or low-resolution satellite data. WorldView-3, Pleiades-HR, and KOMPSAT-3, for example, have swath widths of

13.1, 20, and 16.8 km, respectively. Note that mid-/low-resolution satellite data have dozens or hundreds of kilometers of swath width. These high-resolution satellite cameras frequently use a combination of shorter CCD (Charge-Coupled Device) lines with a slight overlap to increase the swath width [2–6]. As examples, IKONOS, Quickbird, KOMPSAT-3 have three, six, and two overlapping PAN CCD lines, respectively, with shifts in the CCD lines in the scan direction. The merge of each sub-scene from CCD lines is carried out with precise camera calibration information. Each sub-scene is processed considering the sensor alignment, ephemeris effects, and terrain elevations to be merged for a single scene covering a larger swath [2,5].

After the sub-scene merging process, high-resolution satellite data are provided in different processing levels. For example, Maxar provides WorldView data in system-ready, view-ready, and map-ready categories. System-ready imagery allows users to perform custom photogrammetric processes such as digital surface model (DSM) generation and orthorectification using the custom data. View-ready imagery data are products already photogrammetrically processed and designed for users interested in remote sensing applications. Map-ready is a base map that has been orthomosaicked. Level 1R and 1G KOMPSAT-3 data from the Korea Aerospace Research Institute are also available. Level 1R is a product that has been corrected for radiometric and sensor distortions. Level 1G is the product corrected for geometric distortions, including optical distortions and terrain effects, and finally projected to a universal transverse mercator coordinate system.

Many satellite data, including WorldView System-ready and KOMPSAT-3 products, are usually delivered in a single image. This is true when the target area is small enough to be located in an archived image region or a new collection less than the swath width is requested. However, in some cases where the area of interest is large and located crossing over the archived images, users are delivered with a bundle of satellite images. Then, the users have to carry out a photogrammetric process for each data bundle to meet their application purposes.

Typical photogrammetric processes with the bundle of images delivered include orthorectification and mosaics to create a complete image map. The orthorectification requires accurate sensor modeling information such as physical model or rational polynomial coefficients (RPCs) and DSM of the target area. In advance of the orthorectification and mosaic, users should carry out bias compensation of the original sensor model information using ground controls to meet mapping requirements [7]. Then, each image is orthorectified for the DSM and the resulting orthoimages are mosaicked for an image map.

There have been many studies for high-resolution satellite image mosaics in the ground coordinates [8–12]. The proposed algorithms deal with radiometric differences in images caused by seasonal changes [8], image registration and cloud detection with removal [9,10], efficient processing [11], and color balancing [12,13]. Most studies are carried out with photogrammetrically processed orthoimages. However, the cost of these photogrammetric processes should increase with the number of images in the delivered bundle.

With a mosaicked image at the sensor level delivered only with radiometric correction, users should take advantage of more efficient and convenient photogrammetric data processing and management for the simplified data. However, no relevant work on the sensor-level image mosaic was carried out before a photogrammetric process. Firstly, if users are delivered with a single image with single sensor model information instead of multiple data sets, the sensor modeling processing burden should be lifted. This is because users do not have to identify the ground control points on the multiple images. In addition, the tie point extraction process over multiple images is not required for accurate co-registration between the images. Secondly, the orthorectification and mosaic process is simplified because the single image orthorectification is simpler, and mosaic methods, including the seamline generation, are not required.

Therefore, we propose a sensor-level mosaic to generate a seamless image product with geometric accuracy to meet mapping requirements. The approach is different than the ground-level mosaic, as depicted in Figure 1. The ground-level mosaic is carried out with the orthorectification of each image strip to the ground, followed by the seamline extraction and mosaic. As a result, each pixel in the mosaicked image is assigned with map coordinates. In contrast, in the sensor-level mosaic, each image is projected into a reference sensor plane to be merged. The resulting image has single sensor modeling information to relate each mosaic image to the ground.

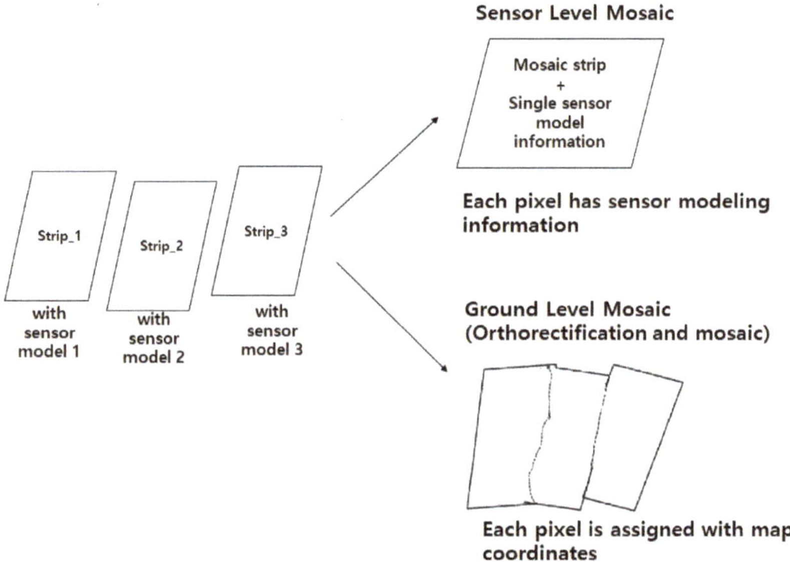

Figure 1. Sensor-level mosaic vs. ground-level mosaic.

The proposed method begins with setting one image to the reference. Each pixel of the other images is projected to the ground using their sensor model information and SRTM (Shuttle Radar Topography Mission) [14] and then projected into the reference using the reference sensor model information. The problem is that the sensor model information is erroneous such that a large geometric discrepancy occurs due to the satellite's inaccurate position and attitude information. Therefore, we aimed to model and remove the irregular difference along the image line using the image matching and outlier removal in the overlapped area.

The paper is structured as follows. In Section 2, the methodology is described with the flowchart with RPCs as the sensor model for image projections. In Section 3, the experimental results are presented for three KOMPSAT-3 strips. The conclusion is presented in Section 4.

2. Methods

The flowchart of the study is given in Figure 2. Given partially overlapped multiple image strips (n images in the figure) and sensor models covering the area of interest, if one image partially overlapped with other images, it was chosen as the reference image. Each pixel of the other images (collateral images) was first projected to the ground using SRTM DEM and then back-projected onto the reference image space. These projections produce $(n - 1)$ projected images partially overlapped with the reference image. Next, image matching was carried out to extract tie points in the overlap area. A lot of matching outliers should exist because of radiometric and geometric differences, such that it requires detecting and remove them accurately. The discrepancy is expected to show irregular patterns along the image line because of push-broom sensor characteristics. Each line of image has a different position and attitude information. Therefore, we modeled the discrepancy with polynomials after dividing the whole image strip into multiple sub-image regions. Based on the polynomial model, outliers are detected and removed in each sub-image region. This leads to the outlier suppressed tie points set, which enables the irregular discrepancy estimation. The mosaicked image strip can be generated after compensating for the image line discrepancy. Finally, single sensor model information for the mosaic image strip is generated.

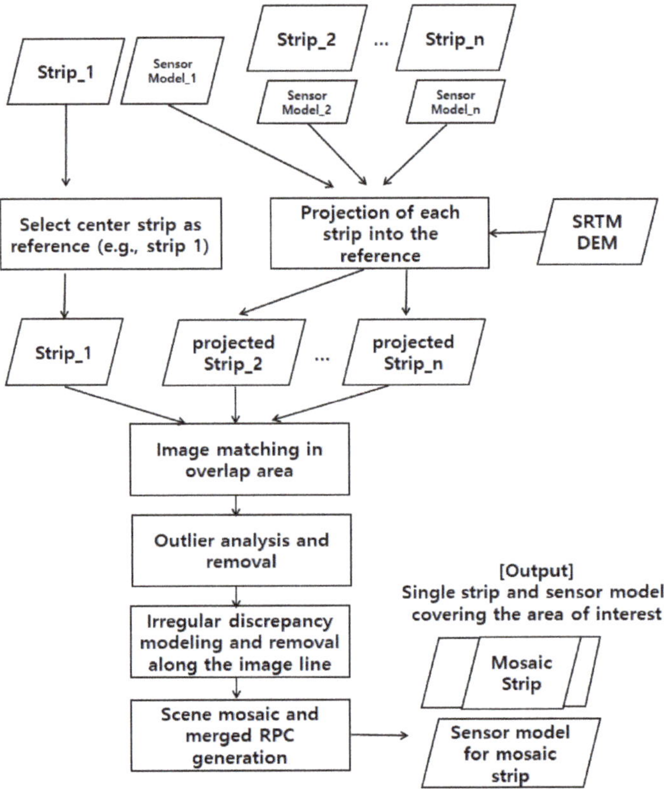

Figure 2. Flowchart of the proposed method for the sensor-level mosaic.

2.1. Projection onto the Reference Image

Except for the reference image, the other images, i.e., collateral images, are required to be projected onto the reference image space using the sensor modeling information. This study used RPCs instead of the physical model for compatibility with little difference in accuracy [15].

1. Ground to image projection:

Ground to image projection is called the forward projection, which equation is expressed as Equation (1). Given 3D ground coordinates (ϕ, λ, h), the corresponding image coordinates (l, s) can be obtained based on the non-linear equation of 78 coefficients (RPCs) [16].

$$
\begin{aligned}
Y &= \frac{Num_L(U,V,W)}{Den_L(U,V,W)} = \frac{a^T u}{b^T u} \\
X &= \frac{Num_s(U,V,W)}{Den_s(U,V,W)} = \frac{c^T u}{d^T u}
\end{aligned}
\tag{1}
$$

with

$$U = \frac{\phi - \phi_O}{\phi_S}, V = \frac{\lambda - \lambda_O}{\lambda_S}, W = \frac{h - h_O}{h_S}, Y = \frac{l - L_O}{L_S}, X = \frac{s - S_O}{S_S}$$
$$u = [1\ V\ U\ W\ VU\ VW\ UW\ V^2\ \ U^2\ W^2\ UVW\ V^3\ VU^2\ VW^2\ V^2U$$
$$U^3\ UW^2\ V^2W\ U^2W\ W^3]^T$$
$$a = [a_1\ a_2\ \ldots\ a_{20}]^T, b = [1\ b_2\ \cdots\ b_{20}]^T, c = [c_1\ c_2\ \cdots\ c_{20}]^T, d = [1\ d_2\ \cdots\ d_{20}]^T$$

where (ϕ, λ, h) are the geodetic latitude, longitude, and ellipsoidal height. (l, s) are the image row and column coordinates. (X, Y) and (U, V, W) are the normalized image and ground coordinates, respectively. $(\phi_O, \lambda_O, h_O, S_O, L_O)$ and $(\phi_S, \lambda_S, h_S, S_S, L_S)$ are the offset and scale factors, respectively for the latitude, longitude, height, column, and row.

However, the major problem is that the target elevation must be given, and there is no closed solution for the ground elevation computation. Figure 3 depicts the iterative ground elevation search process is depicted. Given an image point, the first image to ground projection is performed to the reference elevation, such as the mean elevation of RPCs. The computed horizontal coordinates are used to look up the ground elevation in SRTM DEM. Next, the second image to ground projection is tried for the estimated ground elevation. This iterative process continues until the no changes in the computed horizontal coordinates.

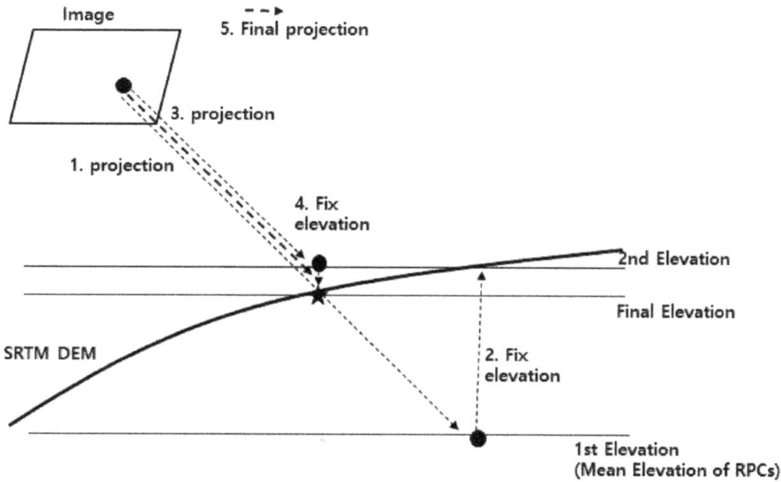

Figure 3. Iterative ground elevation search.

2. Image to ground projection:

Image to ground projection is called the backward projection. Given an image coordinates (l, s) with the ground elevation (h), the horizontal ground coordinates (ϕ, λ) are computed using Equation (2). The backward projection is a non-linear equation that requires to be linearized as Equation (2). The linearized equation requires the initial horizontal ground coordinates (ϕ^0, λ^0) for (U^0, V^0). The solution is obtained by iteration until (dU, dV) it reaches near zero.

$$\begin{bmatrix} V \\ U \end{bmatrix} = \begin{bmatrix} V^0 \\ U^0 \end{bmatrix} + \begin{bmatrix} dV \\ dU \end{bmatrix} \tag{2}$$

$$\begin{bmatrix} dV \\ dU \end{bmatrix} = \begin{bmatrix} \frac{\partial Y}{\partial V}\Big|_{V=V^0} & \frac{\partial Y}{\partial U}\Big|_{U=U^0} \\ \frac{\partial X}{\partial V}\Big|_{V=V^0} & \frac{\partial X}{\partial U}\Big|_{U=U^0} \end{bmatrix}^{-1} \begin{bmatrix} Y - Y^0 \\ X - X^0 \end{bmatrix}$$

where

$$Y^0 = \frac{a^T u^0}{b^T u^0},\ X^0 = \frac{c^T u^0}{d^T u^0}$$
$$u^0 = \left[1\ V^0\ U^0\ W\ V^0U^0\ V^0W\ U^0W\ (V^0)^2\ (U^0)^2\ W^2\ U^0V^0W\ (V^0)^3\ V^0(U^0)^2\ V^0W^2\ (V^0)^2U^0\ (U^0)^3\ U^0W^2\ (V^0)^2W\ (U^0)^2W\ W^3\right]^T$$

$$\frac{\partial Y}{\partial V} = \frac{\partial Y}{\partial u^T}\frac{\partial u}{\partial V},\ \frac{\partial Y}{\partial U} = \frac{\partial Y}{\partial u^T}\frac{\partial u}{\partial U},$$
$$\frac{\partial X}{\partial V} = \frac{\partial X}{\partial u^T}\frac{\partial u}{\partial V},\ \frac{\partial X}{\partial U} = \frac{\partial X}{\partial u^T}\frac{\partial u}{\partial U},$$

2.2. Image Matching and Outlier Removal

Image matching in the overlap area is carried out to extract tie points used for discrepancy compensation. This study uses a template matching based on *NCC* (Normalized Cross-Correlation) as Equation (3). The similarity between reference and projected images is measured using *NCC*. A matching with *NCC* larger than 0.5 is typically considered similar, but a higher threshold such as 0.7 is preferred to reduce matching outliers.

$$NCC = \frac{\sum_{i=1}^{w}\sum_{j=1}^{w}(R_{ij}-\overline{R})(P_{ij}-\overline{P})}{\sqrt{\left[\sum_{i=1}^{w}\sum_{j=1}^{w}(R_{ij}-\overline{R})^2\right]\left[\sum_{i=1}^{w}\sum_{j=1}^{w}(P_{ij}-\overline{P})^2\right]}} \quad (3)$$

where R is a patch in the reference image and P is a patch within the established search region in the projected image, both are in the size of $w \times w$. $\overline{R}, \overline{P}$ are averages of all intensity value in the patches.

These automated image matchings often produce a lot of mismatches that should be detected and removed. RANSAC (Random Sample Consensus) is a popular outlier detection method [17] because it iteratively estimates established modeling parameters from a set of data that includes outliers.

2.3. Piecewise Discrepancy Compensation

High-resolution satellite image strips are acquired using a push-broom sensor that uses a line of detectors arranged perpendicular to the flight direction of the spacecraft. As the spacecraft flies forward, the image is collected one line at a time, with all of the pixels in a line being measured simultaneously.

This mechanism should produce an irregular geometric discrepancy between the adjacent strips along the image line. We applied a piecewise discrepancy compensation that models the local difference for some image lines, as depicted in Figure 4. However, it is a possibility of discontinuity between adjacent image pieces. Therefore, we model each local discrepancy with a spline curve.

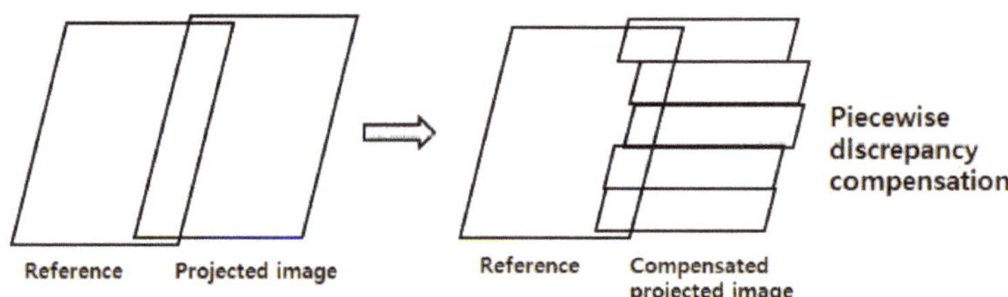

Figure 4. Piecewise discrepancy compensation.

The sensor model for the mosaic image strip should be generated for photogrammetric processes. Since the mosaic image consists of several image strips of different sensor modeling information, the RPCs for the mosaic can be generated by bias-compensating the RPCs of the reference considering the estimated compensations to the adjacent images [14].

3. Experimental Results

3.1. Data

The test data are three image strips of KOMPSAT-3 product level 2R over Romania, as the specifications are listed in Table 1. The acquisition dates are 8 and 24 April and 4 May 2018. The strips have long image line sizes up to 60,000–70,000 pixels with an image swath width of 24,060 pixels. Each image stripe is made up of three image scenes with over 20,000 image lines each. The acquisition geometry includes incidence and azimuth angles. Strips #1 and #3 have similar geometry and a low incidence angle. Small incidence angles of Strips #1 and #3 produce a small GSD (Ground Sample Distance) than Strip #2 with a relatively large incidence angle. Note that the azimuth angle of Strip #2 is in an almost opposite direction from those of the others.

Table 1. Test data specification.

Product Level	Acquisition Date	Image Size (Pixels)	Incidence/Azimuth	GSD (Col/Row)
Strip #1 Level 2R	4 May 2018	Sample 24,060 Line 69,946	0.75°/79.54°	0.55/0.55 m
Strip #2 Level 2R	24 April 2018	Sample 24,060 Line 63,433	26.00°/261.58°	0.67/0.60 m
Strip #3 Level 2R	8 April 2018	Sample 24,060 Line 71,166	11.56°/78.58°	0.56/0.55 m

Figure 5 shows the three data strips. Strip #2 is located in the center with partial overlap with the other strips.

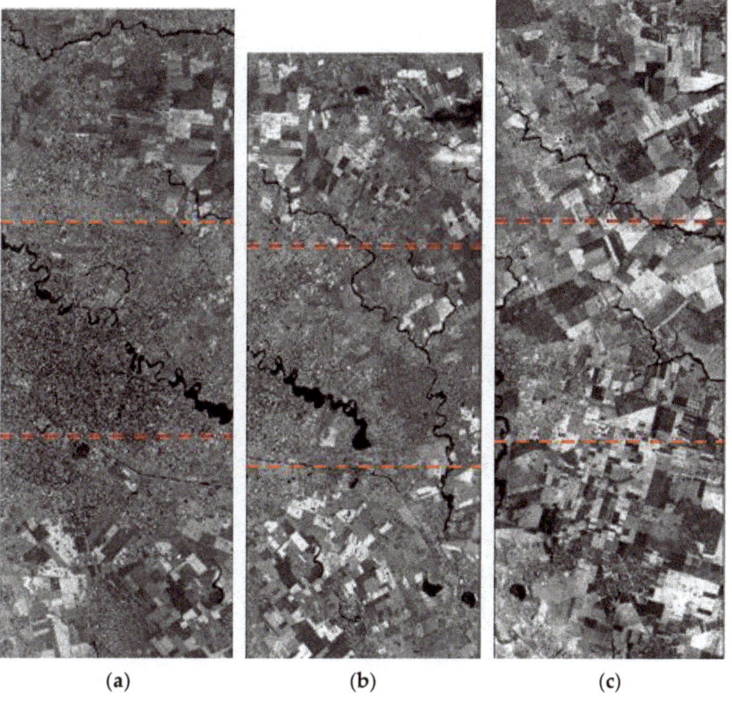

Figure 5. Test image strip of three scenes: (**a**) Strip #1, (**b**) Strip #2, (**c**) Strip #3.

3.2. Sensor Modeling of Each Image Strip

The long strip images were delivered with an ephemeris and attitude data for the physical sensor modeling. However, RPCs are much compatible and easier to use than the physical sensor model, whereas the accuracy is similar. Therefore, we first converted the physical sensor model of each strip into RPCs. The conversion into RPCs is conducted by interpolating satellite attitude information such as roll, pitch, and yaw angles with the first-order equation.

Figure 6 depicts the interpolation residuals for the roll angles of Strip #1, demonstrating that the original roll angle varies locally along the image line. The conversion residuals from the physical model into RPCs are presented in Table 2 for two cases using the original ephemeris and the interpolated ephemeris. Using the interpolated ephemeris shows residuals that are a little better than the other case, which is affected by the local variation in the ephemeris. In Strip #1, the residual in the sample direction improved by more than one pixel.

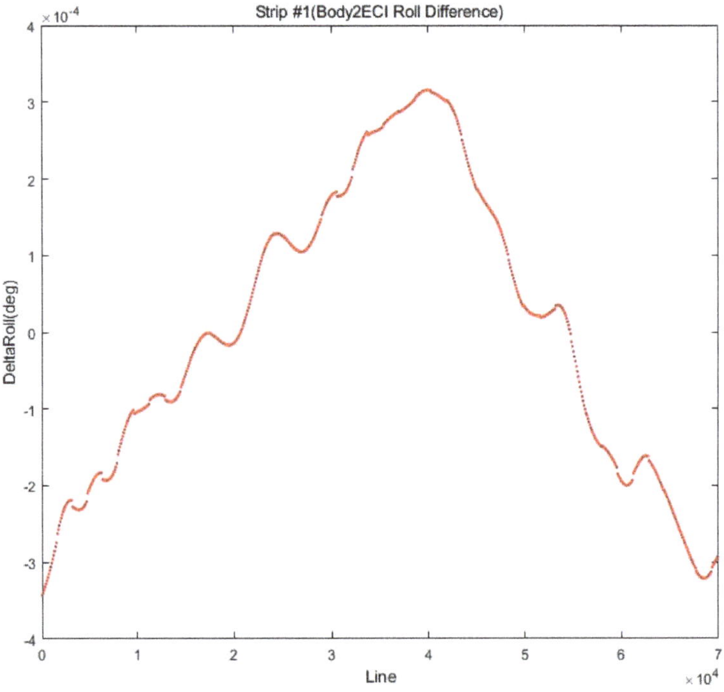

Figure 6. Difference between the original and the interpolated roll angles (Strip #1).

Table 2. RPCs conversion residual in RMSE (unit: pixels).

Ephemeris	Strip #1		Strip #2		Strip #3	
	Line	Sample	Line	Sample	Line	Sample
Original	0.57	1.45	0.55	0.20	0.20	0.28
Interpolated	0.11	0.10	0.10	0.07	0.20	0.10

3.3. Projection of Each Image onto the Reference

We set the center strip (Strip #2) as the reference. Then, we projected each image onto the reference image space using the generated RPCs with 1 arcsec SRTM DEM. First, the reference image is extended to the sides for the image resampling. A point in the extended reference image space is projected iteratively projected onto SRTM DEM as explained in Figure 3, followed by ground to image projection to look up the corresponding digital number in the adjacent strips. Figure 7 depicts three overlaid stripes side by side.

Figure 7. Projected images onto the reference image space.

3.4. Image Matching and Outlier Removal in an Overlap Area

We generated a grid of 50 and 100 pixels along line and sample directions in the overlap area, respectively. Then, we carried out *NCC* image matching between the reference and the adjacent projected images for the grid points. As matching parameters, we used 77×77 pixels for the matching window size, search range 60 pixels. We selected the matching parameters considering the geolocation accuracy of the sensor modeling for KOMPSAT-3, which has 48.5 m (CE90, Circular Error 90% confidence range).

The matching pairs are showing *NCC* larger than 0.7 were selected as matching candidates in this study. Then, the image coordinates differences were computed between the matching pairs and plotted in Figure 8. Figure 8a,b shows the line and sample coordinates differences between Strips #1 and #2. Figure 8c,d shows the line and sample coordinates differences between Strips #2 and #3. The blue dots show all the coordinates differences for the matching candidates.

We applied the RANSAC algorithm with second polynomial models for each line and sample coordinate differences to suppress the matching outliers. The polynomial model was applied to each scene in an image strip. The red dots show the results after the outlier removal.

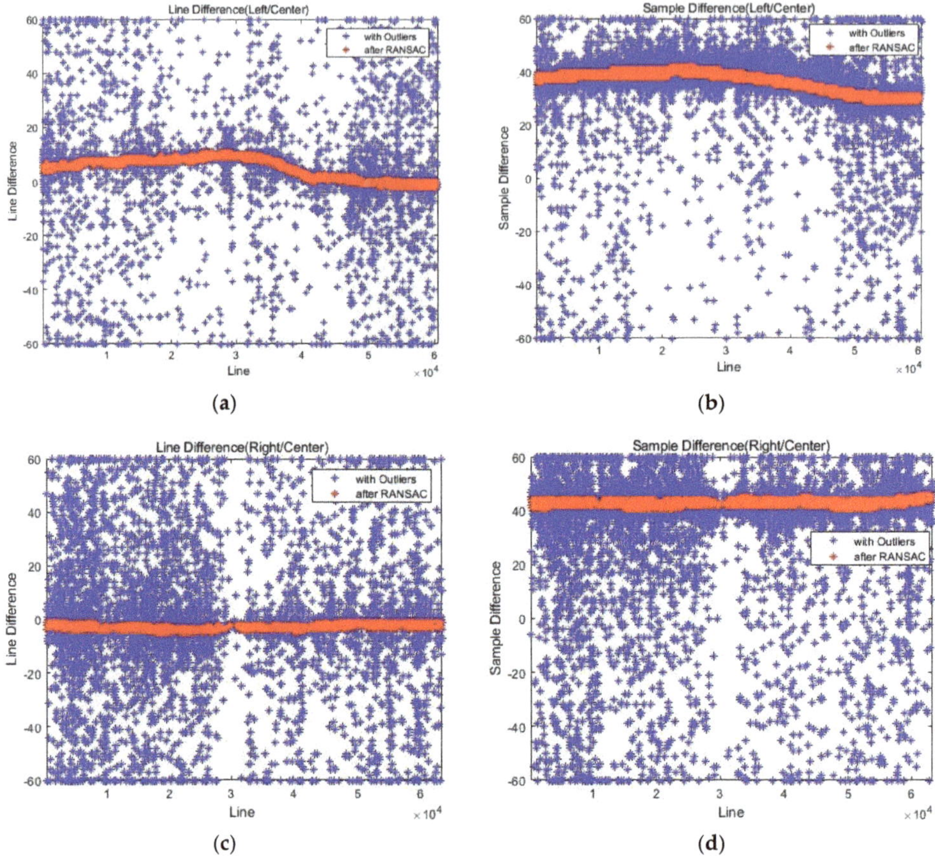

Figure 8. Discrepancy between the image coordinates in the matching pair—(**a**) line difference between Strips #1 and #2; (**b**) sample difference between Strips #1 and #2; (**c**) line difference between Strips #2 and #3; (**d**) sample difference between Strips #2 and #3.

3.5. Piecewise Discrepancy Compensation

After removing the matching outliers, we can estimate the discrepancy compensation of the projected image by averaging the image coordinates differences between the matching pairs. However, the discrepancy varies for each image line. As shown in Figure 7, averaging single image line discrepancies may produce inaccurate compensation values because there are no redundant matching pairs in an image line. Therefore, we estimated the local discrepancy compensation in the line and sample directions by averaging discrepancies in a block of image lines such as 500 image lines. In addition, we interpolated the averaged differences using a spline curve along the image line to ensure the continuity between compensated image blocks.

Figure 9 shows the estimated local discrepancy for the line and sample directions for every 500 image lines after the spline interpolation. In other words, the red line was derived by averaging the red dots in Figure 8 for every 500 image lines and interpolating them in the spline curve. Figure 9a,b shows the line and sample compensations for Strip #1, and Figure 9c,d are for Strip #3. The rewards for sample coordinates ranging from 30 to 44 pixels are much larger than those for line coordinates.

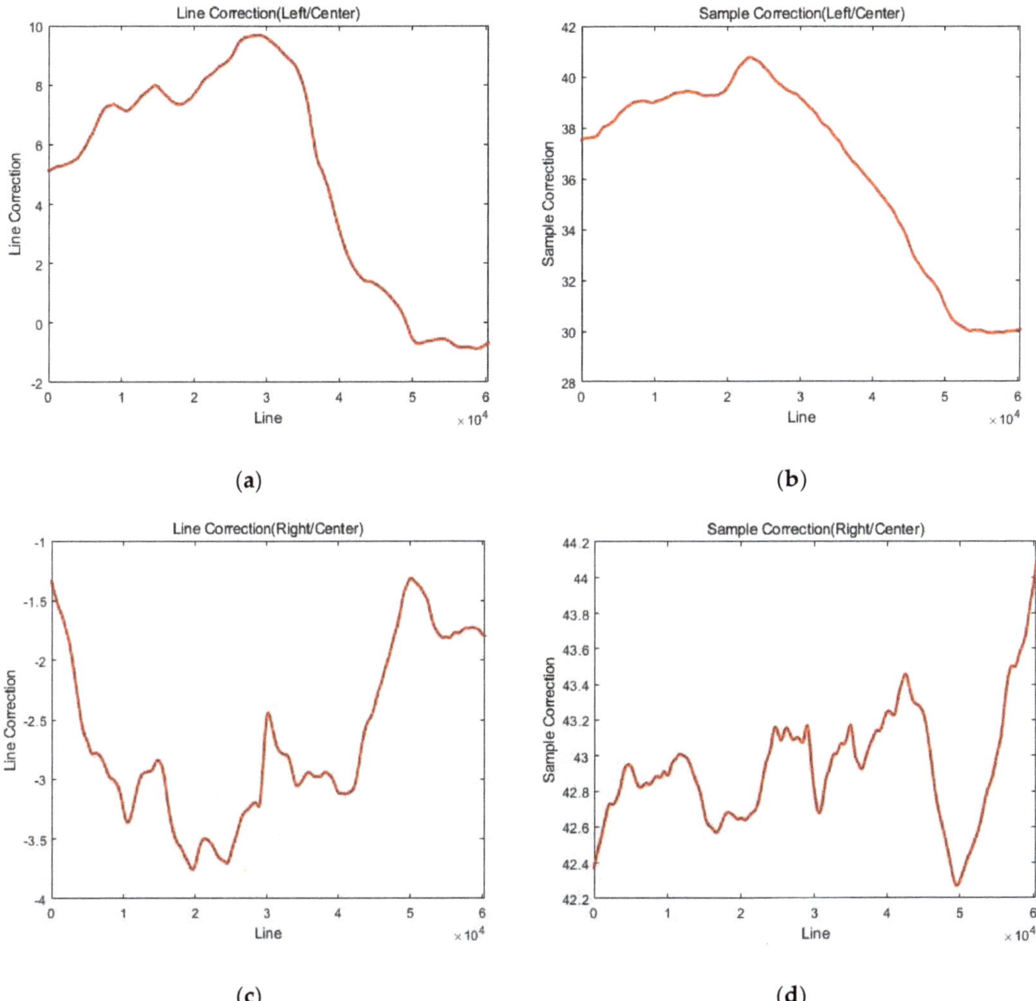

(a) (b)

(c) (d)

Figure 9. Estimated discrepancy compensation—(**a**) line compensation for Strip #1; (**b**) sample compensation for Strip #1; (**c**) line compensation for Strip #3; (**d**) compensation for Strip #3.

The piecewise image compensation produced the final strip mosaic in Figure 10. Note that the color balancing was not carried out in the study. Some examples showing geometric consistency at the strip boundary even over the building areas are presented in Figure 11.

Figure 10. Final strip mosaic.

Figure 11. *Cont.*

Figure 11. Sample images showing geometric consistency at a boundary.

3.6. Sensor Model Information Generation

As the sensor-level strip mosaic was completed, the sensor modeling information for the single mosaic strip was generated for the photogrammetric process. A $7 \times 7 \times 7$ cubic grid covering the whole mosaic image strips was developed in the ground, and the grid points were projected onto the mosaic strip for the corresponding image coordinates. First, only RPCs of the center strip (Strip #2) were extended to cover the whole mosaic image boundary. Secondly, three RPCs were processed together to generate ground and image coordinate sets for single RPCs generation.

To check the accuracy of the generated RPCs, we collected 25 GCPs over the mosaic strip from Google Earth, as shown in Figure 12. We used Google Earth Pro to extract the horizontal and vertical coordinates. Though the accuracy of Google Earth may differ depending on the areas, a few meters of positional accuracy was reported over near urban areas in Europe [18]. First, using the 25 GCPs as checkpoints, we estimated the accuracy of the aforementioned two RPCs of the center strip and mosaic strip, as shown in Table 3. RPCs of the center strip showed rather low positional accuracy of 4.02 and 40.07 pixels in RMSE for the line and sample directions, respectively. However, the RPCs of the mosaic showed much better results reported as 2.88 and 21.07 pixels in RMSE for line and sample directions. The accuracy improvement ranged from 18% to 47.4%. The geolocation performance of the resulted mosaic RPCs seemed improved due to the averaging effects of all RPCs of input data. The RPCs of the mosaic should be more accurate than the RPCs of each strip if more image strips are used for the mosaic.

Table 3. Accuracy of mosaic strip RPCs (unit: pixels).

	RMSE		Max Errors	
	Line	**Sample**	**Line**	**Sample**
RPCs of center strip	4.02	40.07	6.77	45.91
RPCs of Mosaic	2.88	21.07	5.51.	26.78
Accuracy Improvement (%)	28.4%	47.4%	18.6%	41.7%

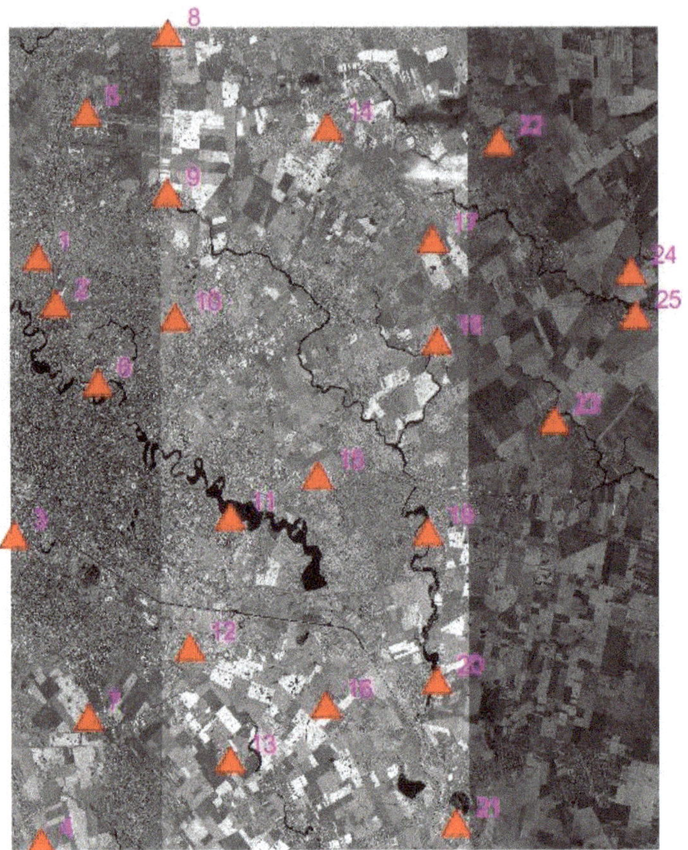

Figure 12. GCP distribution with the number.

Next, the bias compensation of the mosaic RPCs was carried out with the GCPs, and the improved accuracy was presented in Table 4. The bias compensation is a process to improve the input sensor modeling using ground controls. The biases are estimated in image coordinates using the rules and compensated for better accuracy [7]. The errors of the mosaic RPCs were compensated for line and sample directions with constant values estimated from the GCPs. Table 4 shows the RPCs' accuracy after the compensation process. The compensated RPCs showed adequate accuracies ranging from 1.4 to 3.3 pixels in RMSE compared to the ones shown in Table 3.

Table 4. Accuracy of mosaic strip RPCs after the bias compensation (unit: pixels).

	RMSE		**Max Errors**	
	Line	**Sample**	**Line**	**Sample**
Shift	1.44	3.22	3.01	5.96
Linear	1.46	2.79	3.23	5.79

4. Discussion

In the study, we used RPCs instead of rigorous sensor modeling. This is for easier and efficient processing as well as compatibility. However, satellite image providers may use the same approach with their physical sensor model. Regarding image matching, the matching window size and search area can be better optimized considering the area of interest and satellite data specification. For example, fewer features would require a larger matching window size, and satellites with precise sensor models would require a smaller search area. In addition, feature-based image matching methods can be used instead [19]. The discrepancy patterns between image strips in line and sample coordinates would be different for satellite data. The data with stable ephemeris would show rather regular discrepancy patterns along the image lines. However, in any case, image compensation should not be carried out for each image line because there are no redundant matching pairs on a single image line. The sensor modeling of the mosaic tends to be more accurate compared to each image strip due to the averaging effects. Therefore, a mosaic of more image strips would produce better positional accuracy [20].

As shown in the resulting mosaic, the three strips' radiometric differences are observed due to the differences in the acquisition date and angles. The focus of the study is on minimizing the geometric discrepancy and the generation of single sensor model information. Therefore, we have not treated the radiometry in this study, and future research will include the sensor-level radiometric adjustment between the input image strips.

Note that the proposed method is different from the conventional image mosaic carried out with orthorectified images. The proposed sensor-level mosaic is carried out before the photogrammetric processes, including the sensor modeling and orthorectification. Therefore, users can perform their photogrammetric function with the mosaic and the sensor model information.

5. Conclusions

High-resolution satellite images show relatively narrow swath widths such that users often face obstacles to orthorectify and mosaic a bundle of delivered images to create a complete image map. Therefore, the proposed sensor-level mosaicking can generate a seamless image product with improved geometric accuracy. The experimental result with KOMPSAT-3 data showed that the irregular discrepancy between the input images due to the differences in acquisition angles could be minimized for geometrical continuity in the resulted mosaic image. In addition, single sensor modeling information of the mosaic image could be generated for the later photogrammetric processes. The accuracy improvement of the sensor modeling ranged from 18% to 47.4%. Therefore, we believe that the proposed sensor-level mosaic method enables users to take advantage of more efficient and convenient photogrammetric data processing.

Author Contributions: Conceptualization, C.L.; data curation, C.L.; formal analysis, C.L. and J.O.; methodology, C.L. and J.O.; validation, C.L. and J.O.; writing—original draft, J.O.; writing—review and editing, C.L. and J.O. All authors have read and agreed to the published version of the manuscript.

Funding: This study was supported by National Research Foundation of Korea, grant number 2019R1I1A3A01062109.

Institutional Review Board Statement: Not applicable.

Informed Consent Statement: Not applicable.

Data Availability Statement: Not applicable.

Conflicts of Interest: The authors declare no conflict of interest. The funders had no role in the design of the study; in the collection, analyses, or interpretation of data; in the writing of the manuscript, or in the decision to publish the results.

References

1. Loghin, A.M.; Otepka-Schremmer, J.; Pfeifer, N. Potential of pléiades and worldview-3 tri-stereo dsms to represent heights of small isolated objects. *Sensors* **2020**, *20*, 2695. [CrossRef] [PubMed]
2. Seo, D.C.; Lee, C.N.; Oh, J.H. Merge of sub-images from two PAN CCD lines of KOMPSAT-3 AEISS. *KSCE J. Civ. Eng.* **2016**, *20*, 863–872. [CrossRef]

3. Pan, H.; Zhang, G.; Tang, X.; Li, D.; Zhu, X.; Zhou, P.; Jiang, Y. Basic products of the ZiYuan-3 satellite and accuracy evaluation. *Photogramm. Eng. Remote Sens.* **2013**, *79*, 1131–1145. [CrossRef]
4. Cheng, Y.; Jin, S.; Wang, M.; Zhu, Y.; Dong, Z. Image mosaicking approach for a double-camera system in the GaoFen2 optical remote sensing satellite based on the big virtual camera. *Sensors* **2017**, *17*, 1441. [CrossRef] [PubMed]
5. Jacobsen, K. Calibration of imaging satellite sensors. In Proceedings of the International Archives of Photogrammetry and Remote Sensing and Spatial Information Sciences, XXXVI-1/W41, Ankara, Turkey, 14–16 February 2006.
6. Radhadevi, P.V. Pass processing of IRS-1C/1D PAN subscene blocks. *ISPRS J. Photogramm.* **1999**, *54*, 289–297. [CrossRef]
7. Fraser, C.S.; Hanley, H.B. Bias-compensated RPCs for sensor orientation of high-resolution satellite imagery. *Photogramm. Eng. Remote Sens.* **2005**, *71*, 909–915. [CrossRef]
8. Choi, J.; Jung, H.S.; Yun, S.H. An efficient mosaic algorithm considering seasonal variation: Application to KOMPSAT-2 satellite images. *Sensors* **2015**, *15*, 5649–5665. [CrossRef] [PubMed]
9. Zhang, W.; Li, X.; Yu, J. Remote sensing image mosaic technology based on SURF algorithm in agriculture. *J. Image Video Proc.* **2018**, *2018*, 85. [CrossRef]
10. Li, X.; Li, Z.; Feng, R.; Luo, S.; Zhang, C.; Jiang, M.; Shen, H. Generating high-quality and high-resolution seamless satellite imagery for large-scale urban regions. *Remote Sens.* **2020**, *12*, 81. [CrossRef]
11. Chen, H.; He, H.; Xiao, H.; Huang, J. A fast and automatic mosaic method for high-resolution satellite images. In Proceedings of the SPIE 9808, International Conference on Intelligent Earth Observing and Applications, Guilin, China, 9 December 2015. [CrossRef]
12. Cresson, R.; Saint-Geours, N. Natural color satellite image mosaicking using quadratic programming in decorrelated color space. *IEEE J. Sel. Top. Appl. Earth Obs. Remote Sens.* **2015**, *8*, 4151–4162. [CrossRef]
13. Wegmueller, S.A.; Leach, N.R.; Townsend, P.A. LOESS radiometric correction for contiguous scenes (LORACCS): Improving the consistency of radiometry in high-resolution satellite image mosaics. *Int. J. Appl. Earth Obs. Geoinf.* **2021**, *97*, 102290. [CrossRef]
14. Farr, T.G.; Rosen, P.A.; Caro, E.; Crippen, R.; Duran, R.; Hensley, S.; Kobrick, M.; Paller, M.; Rodriguez, E.; Roth, L.; et al. The shuttle radar topography mission. *Rev. Geophys.* **2007**, *45*. [CrossRef]
15. Dial, G.; Grodecki, J. RPC Replacement Camera Models. In Proceedings of the American Society for Photogrammetry and Remote Sensing 2005 Annual Conference, Baltimore, MD, USA, 7–11 March 2005.
16. Grodecki, J. IKONOS stereo feature extraction—RPC approach. In Proceedings of the ASPRS 2001, St. Louis, MO, USA, 23–27 April 2001.
17. Fischler, M.A.; Bolles, R.C. Random sample consensus: A paradigm for model fitting with applications to image analysis and automated cartography. *Commun. ACM* **1981**, *24*, 381–395. [CrossRef]
18. Pulighe, G.; Baiocchi, V.; Lupia, F. Horizontal accuracy assessment of very high resolution Google Earth images in the city of Rome, Italy. *Int. J. Digit. Earth.* **2016**, *9*, 342–362. [CrossRef]
19. Oh, J.; Han, Y. A double epipolar resampling approach to reliable conjugate point extraction for accurate Kompsat-3/3A stereo data processing. *Remote Sens.* **2020**, *12*, 2940. [CrossRef]
20. Rottensteiner, F.; Weser, T.; Lewis, A.; Fraser, C.S. A strip adjustment approach for precise georeferencing of ALOS optical imagery. *IEEE Trans. Geosci. Remote Sens.* **2009**, *47*, 4083–4091. [CrossRef]

Article

Coupling Denoising to Detection for SAR Imagery

Sujin Shin *, Youngjung Kim, Insu Hwang, Junhee Kim and Sungho Kim

Agency for Defense Development, Institute of Defense Advanced Technology Research, Daejeon 34186, Korea; read12300@add.re.kr (Y.K.); hciinsu@add.re.kr (I.H.); kjh1127@add.re.kr (J.K.); cocktail@add.re.kr (S.K.)
* Correspondence: sujinshin@add.re.kr; Tel.: +82-42-821-4639

Featured Application: The proposed object detection framework aims to improve detection performance for noisy SAR images, which is applicable for general object detection in SAR imagery: recognition of militarily important targets such as ships and aircrafts or monitoring for abnormal civilian events.

Abstract: Detecting objects in synthetic aperture radar (SAR) imagery has received much attention in recent years since SAR can operate in all-weather and day-and-night conditions. Due to the prosperity and development of convolutional neural networks (CNNs), many previous methodologies have been proposed for SAR object detection. In spite of the advance, existing detection networks still have limitations in boosting detection performance because of inherently noisy characteristics in SAR imagery; hence, separate preprocessing step such as denoising (despeckling) is required before utilizing the SAR images for deep learning. However, inappropriate denoising techniques might cause detailed information loss and even proper denoising methods does not always guarantee performance improvement. In this paper, we therefore propose a novel object detection framework that combines unsupervised denoising network into traditional two-stage detection network and leverages a strategy for fusing region proposals extracted from both raw SAR image and synthetically denoised SAR image. Extensive experiments validate the effectiveness of our framework on our own object detection datasets constructed with remote sensing images from TerraSAR-X and COSMO-SkyMed satellites. Extensive experiments validate the effectiveness of our framework on our own object detection datasets constructed with remote sensing images from TerraSAR-X and COSMO-SkyMed satellites. The proposed framework shows better performances when we compared the model with using only noisy SAR images and only denoised SAR images after despeckling under multiple backbone networks.

Keywords: denoising; detection; SAR imagery; fusing region proposals

Citation: Shin, S.; Kim, Y.; Hwang, I.; Kim, J.; Kim, S. Coupling Denoising to Detection for SAR Imagery. *Appl. Sci.* **2021**, *11*, 5569. https://doi.org/10.3390/app11125569

Academic Editor: Yang-Dam Eo

Received: 12 May 2021
Accepted: 8 June 2021
Published: 16 June 2021

1. Introduction

Synthetic Aperture Radar (SAR) is a type of radar system used to reconstruct 2D or 3D terrain and objects on the ground (or over oceans). The SAR system utilizes a technology to synthesize a long virtual aperture through a coherent combination of the received signals from objects. The synthesized aperture transmits pulses of microwave radiation, which in turn has the effect of narrowing the effective beam width in an azimuth direction and thus achieving high resolution. Combining return signals by an on-board radar antenna, SAR overcomes the main limitations of traditional systems that the azimuth resolution is determined by physical antenna size. Optical and infrared sensors are passive since they detect objects by reflected light and emitted signals from the objects, respectively, while the radars can actively transmit and receive radar waves, operating in all-weather and day-and-night conditions.

Thanks to the useful characteristics available under all-weather conditions and also during night-time, SAR images are especially applied to military reconnaissance as most military operations take place at night in poor weather conditions. There is a variety of applications such as information and electronic warfare, target recognition of aircrafts that maneuver irregularly, battlefield situational awareness, and development of aircrafts that are hard for the other party to track with radar. In addition, it is necessary to study on object detection using radar imagery for civilian applications (e.g., resources exploration, environmental monitoring, etc.).

With the recent rapid development of deep learning, many deep convolutional neural network (CNN)-based object detection approaches using SAR imagery have gained increased attention. The successes of the deep detectors on SAR images facilitate a wide range of civil and military applications, such as detection of ship [1–5], aircraft [6–9], destroyed building [10], oceanic internal wave [11], oceanic eddy [12], oil spill [13], avalanche [14], and trough [15]. For the further research purposes, several SAR object detection datasets have also been released called AIR-SARShip-1.0 [16], SAR-Ship-Dataset [17], SAR ship detection dataset (SSDD) [18], and HRSID [19].

SAR images are formed from a coherent sum of backscattered signal components at the boundary of different media after pulsed transmissions of microwave radiation, enabling to observe the interior of the targets otherwise invisible to the naked eye. However, when obtaining the SAR images, if the emitted pulses are reflected from the boundary of a target with uneven surface, then scattering and interference waves are created. These wave signals have a direct impact on a SAR imaging the structure of the target as noise components. The produced noise is often called *speckle noise*, which hinders the original image information and causes a speckle corrupted SAR image as shown in Figure 1. The scattering characterization of the target gets severe depending on changes in radial properties and orbital surfaces, leading to degradation of recognition performance. It is worth noting that a number of published studies were conducted for denoising (or despcekling) SAR images [20–25].

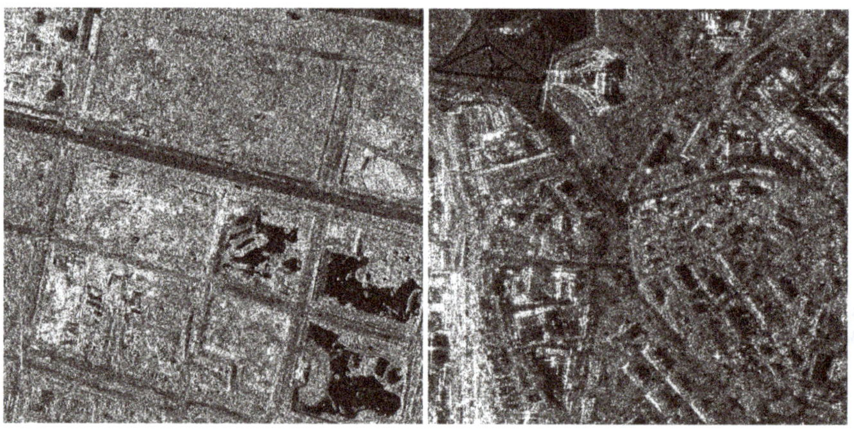

(**a**) TerraSAR-X (**b**) COSMO-SkyMed

Figure 1. Examples of the real-world SAR image where noise-like speckle appears.

Many previous works first perform despeckling on SAR images as one of preprocessing steps and then utilize the SAR images for several tasks via deep learning; e.g., classification task [26,27], detection task [28–30], etc. Processing separately the large amount of SAR images results in high time consumption and low efficiency. Though various despeckling methods such as Lee filter [22], Kuan filter [23], Frost filter [24], Probabilistic Patch-Based (PPB) filter[25] have been proposed, if we take an improper despeckling methodology without considering the dataset characteristics carefully, then the despeckling may lead to poor performance due to the information loss from raw SAR images. Meanwhile, to further improve the visual quality of SAR images, there are other preprocessing methods such as contrast enhancement methods. Given that most of SAR images are usually grayscale images, we can consider various processing methods, for example, fuzzy-based gray-level image contrast enhancement [31] or fuzzy-based image processing algorithm [32].

To overcome the issue and guide for directly promoting object detection performance, developing an object detection framework through incorporating an alternative deep denoiser replacing the separate denoising preprocessing step into the classical object detection network is significant and necessary. The motivation shares the similar spirit to the recent classification work proposed by Wang et al. [33], where they learn a noise matrix from an input noisy image and with the noise matrix synthesize a despeckled image taken as the input into a subsequent classification network. According to our best knowledge, we are the first to connect a denoising network to an object detection network. We additionally introduce *fusing region proposals* approach which fuses set of Region of Interests (RoIs) from both noisy and denoised images; rather than simply ending with the coupling structure as in Wang et al. [33].

We propose a novel object detection framework whose the core idea comprises two parts: (1) connecting an unsupervised denoising network to an object detection network for dynamically extracting a denoised SAR image from a given noisy SAR image, and (2) forwarding an image pair of two SAR images (the given real SAR image and the synthetically generated SAR image) to an object detection network and fusing region proposals from the two SAR images for complementarily integrating regional information. Here *fusing region proposals* refers to merging two sets of RoIs yielded by a shared region proposal network within the object detection network. This is inspired by the observation that utilizing only real SAR image may bring about false positives due to the inherent speckle noise of the image and on the contrary, depending on only denoised SAR image may cause missing targets because inadequate denoising leads to fine information loss of raw data.

The rest of this paper is organized as follows. Section 2 mainly consists of two parts, where the first part introduces our datasets constructed with SAR images from TerraSAR-X and COSMO-SkyMed satellites, and the second part describes the detailed design of our proposed object detection framework, i.e., how to incorporate an unsupervised denoising network into an object detection network and fuse the region proposals within the object detection network. Section 3 reports comparative experimental results for the proposed object detection network on our own datasets. To validate the effectiveness of our approach, we carry out multiple experiments; (1) we need to experimentally demonstrate that our coupling structure between denoising and detection networks can strengthen detection performance, (2) we further verify the proposed region proposal fusing strategy in terms of input data for detection network and fusing method through ablation studies, and (3) we additionally perform comparative experiments with respect to the choice of a feature map extracted from either real or synthetic SAR image, where the feature map refers to the output of CNN backbone in the detection network. Section 4 presents the discussion of the experimental results together with an additional time complexity analysis. Finally, Section 5 includes the final remarks and a conclusion.

2. Materials and Methods

In this section, we describe SAR remote sensing datasets that we constructed and the proposed object detection framework which fuses region proposals utilizing denoised SAR image. The remote sensing datasets include not only SAR imagery but also corresponding labeled objects. We develop our object detection framework with the datasets and detail the proposed framework in the rest of this section.

2.1. SAR Remote Sensing Dataset

2.1.1. Description

We constructed our datasets with 60 TerraSAR-X images from German Aerospace Center [34] and 55 COSMO-SkyMed images from Italian Space Agency [35], which is mainly covering harbor- and airport- peripheral areas. For TerraSAR-X satellite, the images have resolutions from 0.6 m to 1 m, and is of the size in the range from about 6 k × 2 k to 11 k × 6 k pixels (sorted by their area). For COSMO-SkyMed satellite, the images have a resolution of 1m, and is of the size in the range from about 13 k × 14 k to 20 k × 14 k pixels (sorted by their area). Each remote sensing image is labeled by experts in aerial image interpretation with multiple categories such as airplane (A), etcetera (E) and ship (S). The ship/airplane classes contain a variety of civil and military ships/airplanes while the etcetera class includes support vehicles, air defense weapons and air defense vehicles. Some example ship/airplane objects are shown in Figures 2 and 3 for TerraSAR-X and COSMO-SkyMed imagery, respectively.

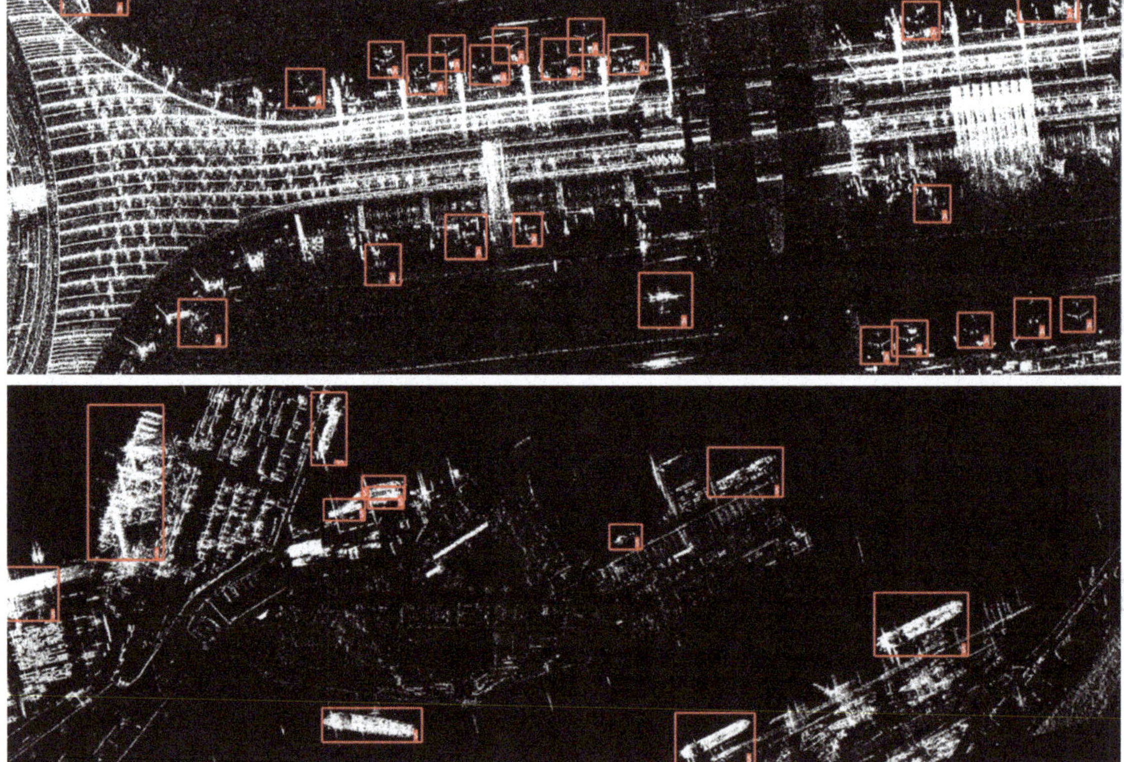

Figure 2. Example airplane (top) and ship (bottom) objects in TerraSAR-X image. The groundtruth bounding boxes labeled as corresponding class are plotted in red color.

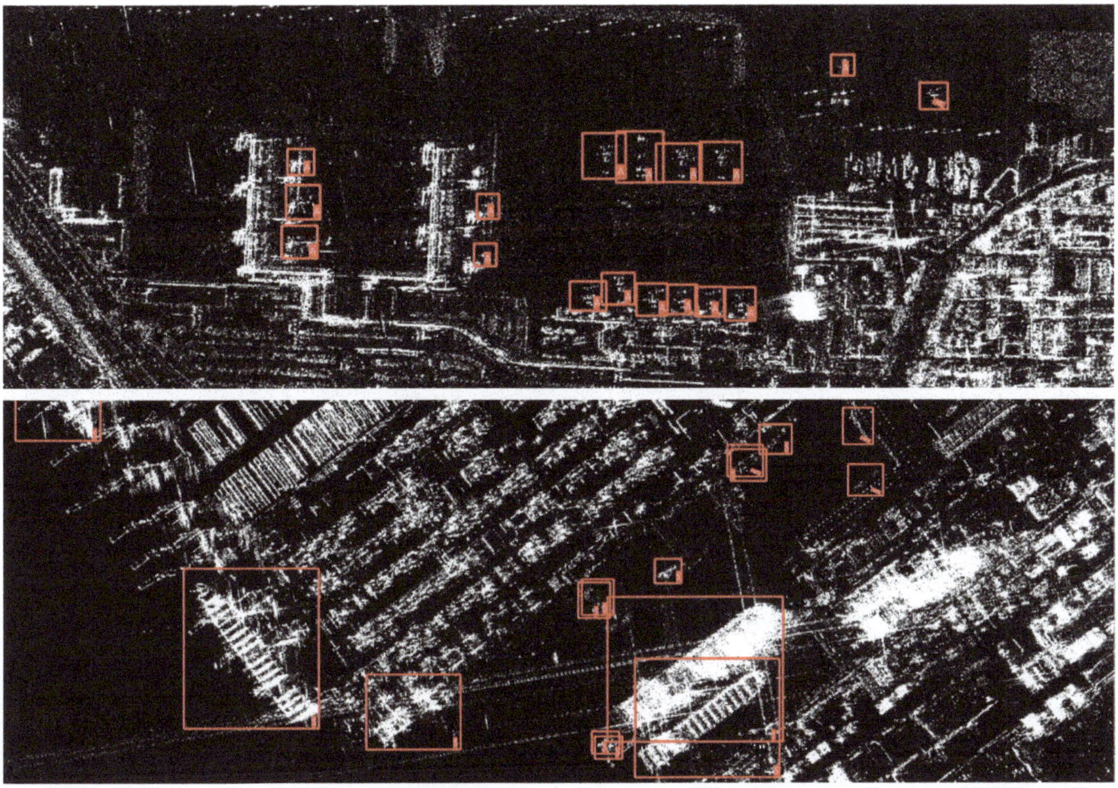

Figure 3. Example airplane (top) and ship (bottom) objects in COSMO-SkyMed image. The groundtruth bounding boxes labeled as corresponding class are plotted in red color.

Our labeled objects include a total of 15.7 k instances of 3 categories; 3.7 k instances for A class, 0.2 k instances for E class, and 11.8 k instances for S class, which implies that our datasets are quite imbalanced between the categories and relatively skewed towards S class. The class distribution by type of satellite imagery is plotted in Figure 4. Furthermore, target objects in our dataset exist at a variety of scales due to our multiresolution images and the variety of shapes, especially for ships objects. We measure the bounding box size of objects with $w_{bbox} \times h_{bbox}$ and present the frequency of boxes by size as a histogram in Figure 5, where w_{bbox} and h_{bbox} is the width and height of the bounding box, respectively.

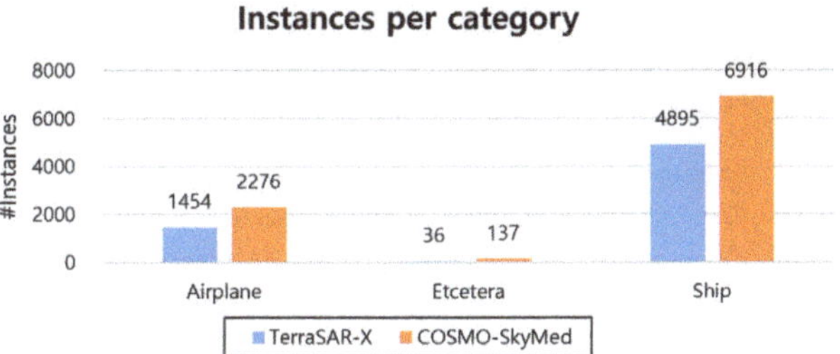

Figure 4. Number of annotated instances per category for TerraSAR-X and COSMO-SkyMed imagery.

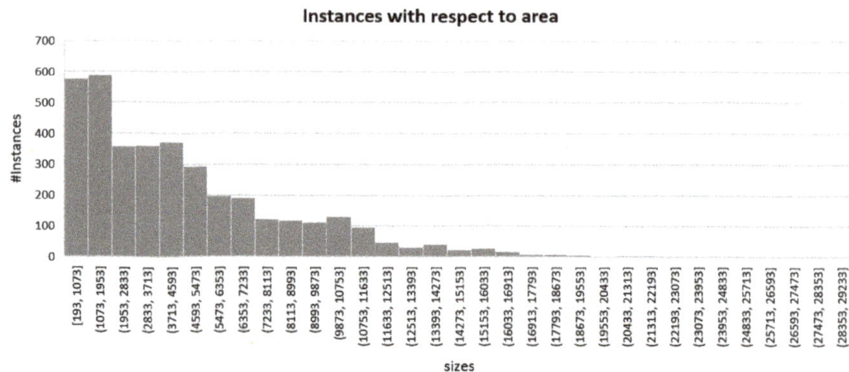

Figure 5. Histogram that exhibits the number of annotated instances with respect to area (width × height) in pixels.

2.1.2. Comparison to other SAR Detection Datasets

Table 1 summarizes the detailed comparisons between our own constructed dataset and other publicly available SAR detection datasets, i.e., AIR-SARShip-1.0 [16], SSDD [18], SAR-Ship-Dataset [17], and HRSID [19]. SAR-Ship-Dataset is the dataset with the largest number of instances, followed by our own dataset. The primary differentiator of our dataset as compared with other datasets lies in (1) class diversity such as ships, aircrafts, and etcetera classes, and (2) the number of scene areas. We obtained the SAR images from a variety of harbor and airport peripheral areas around the world wide and annotated different shapes of objects.

Table 1. Comparison of statistics among multiple datasets. We denote the number of instances, patches, and areas as # Instances, # Patches, and # Areas, respectively.

Dataset	# Instances	# Patches	# Areas	Patch Size	Resolution
AIR-SARShip-1.0 [16]	461	31	4	3000 × 3000	1∼3 m
SSDD [18]	2540	1160	15	300 × 400	1∼10 m
SAR-Ship-Dataset [17]	59,535	43,819	30	256 × 256	3∼25 m
HRSID [19]	16,951	5604	13	800 × 800	0.6∼3 m
Our Dataset	21,717	16,308	92	800 × 800	0.6∼1 m

2.2. Proposed Methodology

Given the inherent speckle noise of SAR, researchers have previously performed a preprocessing step like despeckling before training an object detection model. However, such prior preprocessing independent of the performance of object detection may not only be inefficient, but also lead to weak detection performance because an unintentionally improper denoising induces loss of detailed information. Therefore, we integrate a denoising network with a two-stage detection network so that the denoising network can directly receive feedbacks from the detection network, as illustrated in Figure 6.

We choose a blind-spot neural network [36] based self-supervised scheme as the unsupervised denoising model and adopt Gamma noise modeling as in Speckle2Void [37] fitted with SAR speckle, but not limited to this model sturcture. We can train the unsupervised denoising model as a generator G that maps a real (noisy) SAR image I_{real} to the synthetic (denoised) SAR image $G(I_{real})$. The core idea of our model is to infer a synthetic denoised SAR image from the input SAR image and merge the two sets of extracted RoIs to improve detection performance. Without any help of related materials such as corresponding denoised image for an input SAR image, we can autonomously simulate the denoised image and fuse the inferred information such as RoIs. The entire model enables effective end-to-end learning.

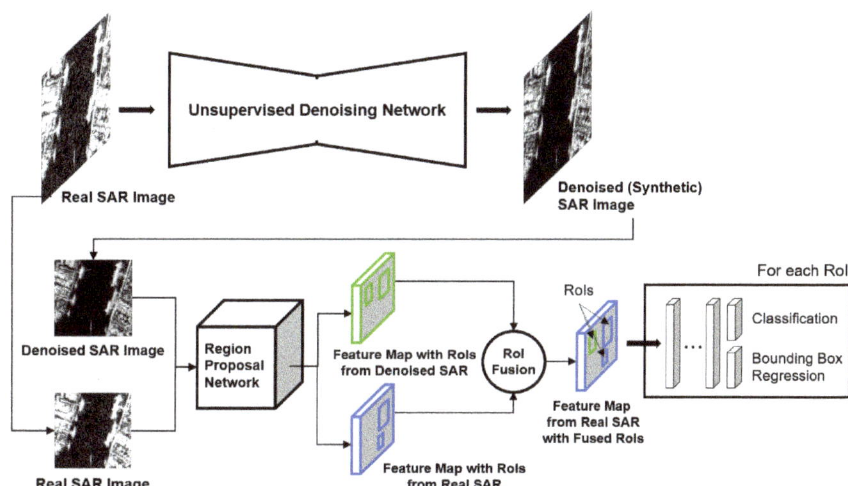

Figure 6. Overview of the proposed object detection framework: (1) connecting an unsupervised denoising network to an object detection network for dynamically extracting a denoised SAR image from a given noisy SAR image, and (2) forwarding an image pair of two SAR images to an object detection network and fusing region proposals from the two SAR images for complementarily integrating regional information.

The unsupervised denoising network G in our model firstly takes as input a real (noisy) SAR image I_{real} and extracts synthetic (denoised) SAR image $G(I_{real})$ as the output. Then, the formed (real, synthetic) image pairs $(I_{real}, G(I_{real}))$ are fed into a shared region proposal network and the region proposal network outputs two corresponding feature maps and sets of RoIs. The two sets of RoIs $\mathbb{B}_{real}, \mathbb{B}_{synth}$ are merged and the redundant bounding boxes are subsequently removed by a NMS procedure, i.e., $\mathbb{B}_{final} = NMS(\mathbb{B}_{real} \cup \mathbb{B}_{synth})$, where $\mathbb{B}_{final}$ is the resultant fused bounding boxes. For each RoI in $\mathbb{B}_{final}$ on the feature map from the real SAR image, the RoI feature vector is then forwarded to obtain the classification and regression results as traditional two-stage detection network.

Usually, only single SAR image which is either real or denoised (preprocessed) is employed for training an object detection network as shown in Figure 7. Suppose we have real SAR images which is inherently speckled noisy without any preprocessing, relying solely on the real SAR image for training may cause false alarms of region proposals. On the other hand, utilizing denoised SAR images alone may be prone to suffer from missing targets because of detailed information loss. We, therefore, devise a novel denoising-based object detection network to make full use of the complementary advantages between the real and denoised SAR images.

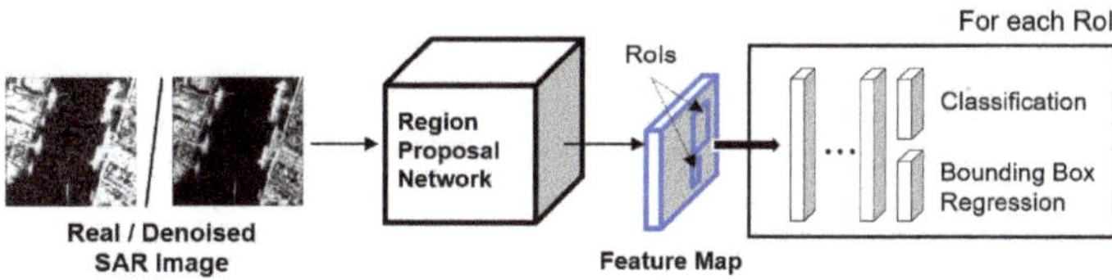

Figure 7. Overview of the traditional two-stage object detection network given a real or denoised (preprocessed) SAR image as input.

To combine extracted information from both real and synthetic SAR images, we consider *fusing region proposals* which merges two sets of RoIs yielded by a region proposal network. Considering that there exist qualitative differences between the two sets of RoIs derived real and synthetic SAR images, the real and synthetic SAR images are separately trained by the region proposal network. After fusing region proposals, we take the feature map from the real SAR image for preserving the global context information of the raw input SAR image.

The proposed architecture is trained end-to-end with a multi-task loss which mainly consists of (1) unsupervised denoising loss, (2) region proposal loss, and (3) RoI loss for classification and bounding-box regression. Especially, the region proposal network is trained for both real and synthetic SAR image, and thus two distinctly losses are defined. The final loss function that we propose is a weighted summation of all losses as follows.

$$\mathcal{L}(I_{real}) = \lambda_1 \mathcal{L}_{den}(I_{real}) + \lambda_2 \mathcal{L}_{rpn}^{real}(I_{real}) + \lambda_3 \mathcal{L}_{rpn}^{synth}(G(I_{real})) + \lambda_4 \mathcal{L}_{roi}(\mathbb{B}_{final}) \quad (1)$$

where:

$$I_{real} = \text{a real (noisy) image}$$
$$G(I_{real}) = \text{a synthetic (denoised) image extracted from the denoising network } G$$
$$\mathbb{B}_{final} = NMS(\mathbb{B}_{real} \cup \mathbb{B}_{synth}), \text{ where } \mathbb{B}. \text{ is set of RoIs from either } I_{real} \text{ or } G(I_{real})$$

where $\mathcal{L}_{den}$ denotes the unsupervised denoising loss. $\mathcal{L}_{rpn}^{real}$ and $\mathcal{L}_{rpn}^{synth}$ are the region proposal loss of RPN for I_{real} and $G(I_{real})$, respectively. $\mathcal{L}_{roi}$ refers to the loss summation of classification and bounding-box regression loss for all RoIs $\mathbb{B}_{final}$. $\lambda_{1:4}$ are the hyper-parameters to balance the interplay between the losses and the all parameters are set to 1 in all our experiments.

3. Results

We first present the description of our experimental dataset settings in Section 3.1. Section 3.2 presents the details of our model architecture and the hyperparameter settings. Based on this implementation, we conduct extensive experiments to validate the contributions of the proposed model and Sections 3.3 and 3.4 contain the experimental results. Section 3.5 provides comprehensive ablation studies.

3.1. Dataset Settings

We acquired 60 TerraSAR-X raw scenes from German Aerospace Center [34] and 55 COSMO-SkyMed raw scenes from Italian Space Agency [35]. The raw scenes go through multiple stages like preprocessing, Doppler centroid estimation (DCE), and focusing to obtain single look slant range complex (SSC) images. The SSC images are then converted to multi-look ground range detected (MGD) images by multi-looking procedures. With the MGD images, we create patches of size 800×800 via sliding-window operation, within each patch containing at least one target object which belongs to airplane (A), etcetera (E), or ship (S) categories. Finally, we randomly split patches into 80% for training, and 20% for testing.

3.2. Implementation Details

We implemented our unsupervised denoising model following self-Poisson Gaussian [38], however, adopted Gamma noise modeling as in Speckle2Void [37] to characterize the SAR speckle. Our implementation for detection framework was based on the MMDetection tool box [39] which is developed in PyTorch [40]. Stochastic gradient descent (SGD) Optimizer [41,42] with momentum of 0.9 was used for optimization. We trained a total of 24 epochs, with an initial learning rate of 0.0025, momentum of 0.9, and weight decay of 0.0001. We experimented with ResNet-50-FPN and ResNet-101-FPN backbones [43,44]. All evaluations were carried out on a TITAN Xp GPUs with 12G memory.

3.3. Qualitative Evaluation

Figure 8 shows paired examples of real SAR images and corresponding synthetically denoised SAR images where the denoised SAR images are the intermediate results in our model. After the denoising stage, the general speckle noises are drastically reduced; however, there inevitably exists a trade-off between the noise level and image clarity. Especially, a lot of buoys that usually look like actual ships are located in the first example of Figure 8 and in the denoised SAR image, brightness of the buoys relatively gets faded and the visual difference with the surrounding ships becomes clear. In addition, scattering waves around target objects which are one of factors hindering accurate localization is blurred after the denoising. The denoising within our network confirms such positive effectiveness.

Some image triples of groundtruth, baseline detection, and our detection visualizations are presented in Figure 9. We train the baseline detection model with non-preprocessed and raw noisy SAR images. For a fair comparison, both the baseline and our detection model equally adopt Faster RCNN with ResNet-101-FPN [43,44] backbone architecture. The detection results show that our model could localize overall objects accurately with higher confidence scores and detects with a small number of false alarms compared to with the baseline detection model in the given patch images. Although the progress made by our detection models are inspiring, our detectors still have a room further improvement due to the few remaining false alarms and missing targets.

3.4. Quantitative Evaluation

To quantitatively evaluate the detection performance, we calculate mean average precision (mAP). The mAP metric is widely used as a standard metric to measure the performance of object detection and estimated as the average value of AP over all categories. Here, AP computes the average value of precision over the interval from recall = 0 to recall = 1. The precision weighs the fraction of detections that are true positives, while the recall measures the fraction of positives that are correctly identified. Hence, the higher the mAP, the better the performance.

As shown in Table 2, we compare the proposed network with the traditional two-stage detection model under two different backbones such as ResNet-50-FPN and ResNet-101-FPN [43,44]. By varying despeckling approaches, we set several baseline models as previous work processes: (1) inputting non-preprocessed real SAR images, (2) feeding denoised SAR images into the traditional two-stage detection model after denoising via representative techniques called Lee filter [22] or PPB filter [25]. We observe that the despeckling effect of applying Lee filter is more minor than PPB filter. PPB filter enables us to reduce more speckle noises; but, much detailed information visually gets concealed. This validates our experimental results that the baseline model with PPB filter slightly performs inferior compared to the baseline model with Lee filter. On the other hand, our detection network provides significant advances in performance under all backbone architectures. Through observation of the test results, this is attributed to the suppression of many false positive detections resulting from speckle noise problems of real SAR images.

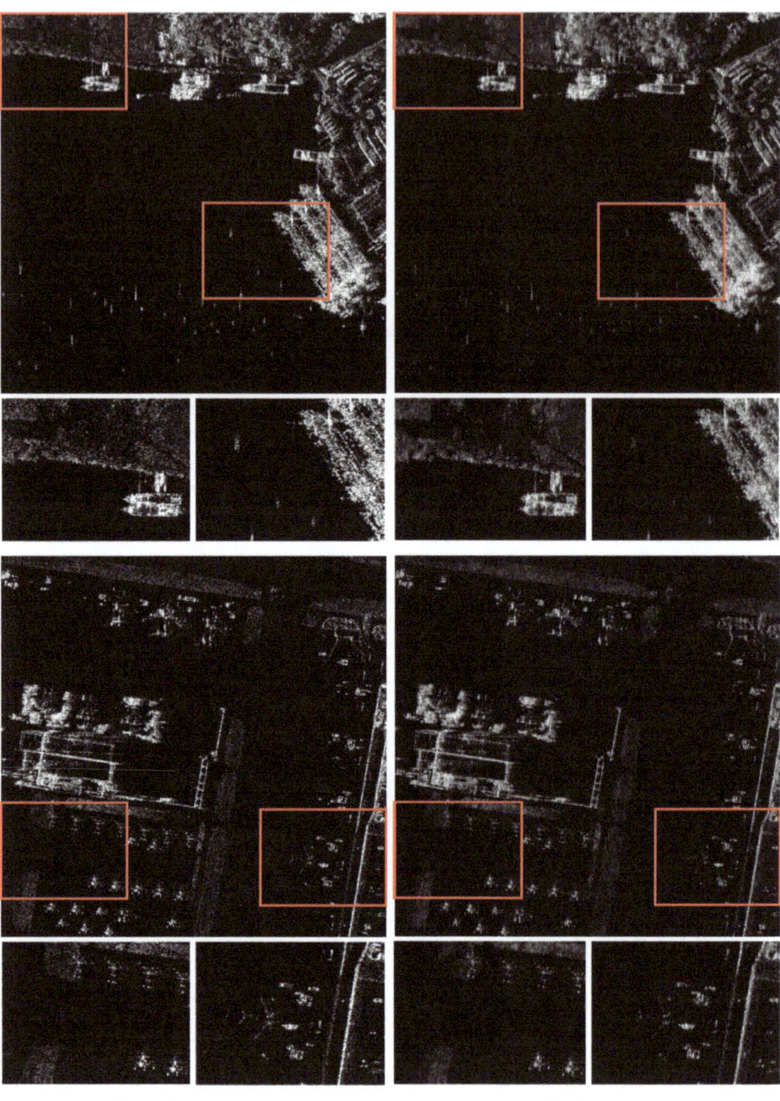

(a) Real SAR image (b) Synthetically denoised SAR image

Figure 8. Two paired examples of noisy SAR (left) and despeckled SAR (right) images. Red bounding boxes for each image enlarge corresponding sub-regions. As shown in the enlarged windows, scattering waves and speckle noises are relatively less observed in denoised examples.

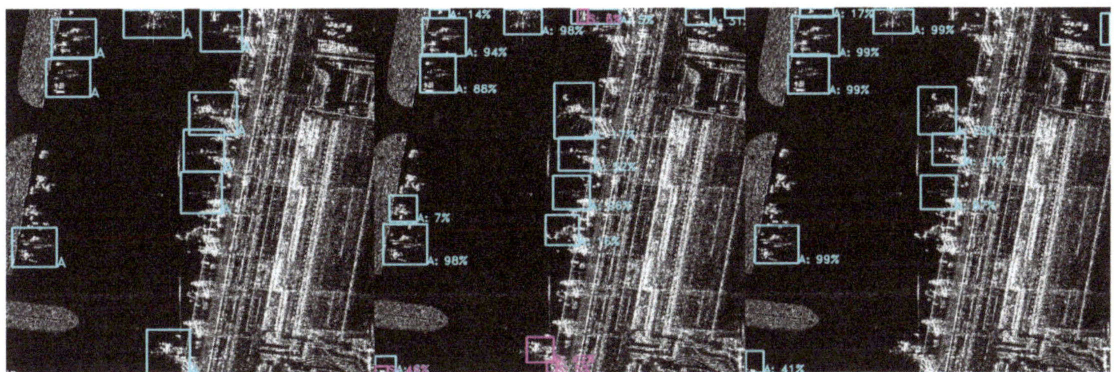

(a) Airplane (A) class

(b) Etcetera (E) class

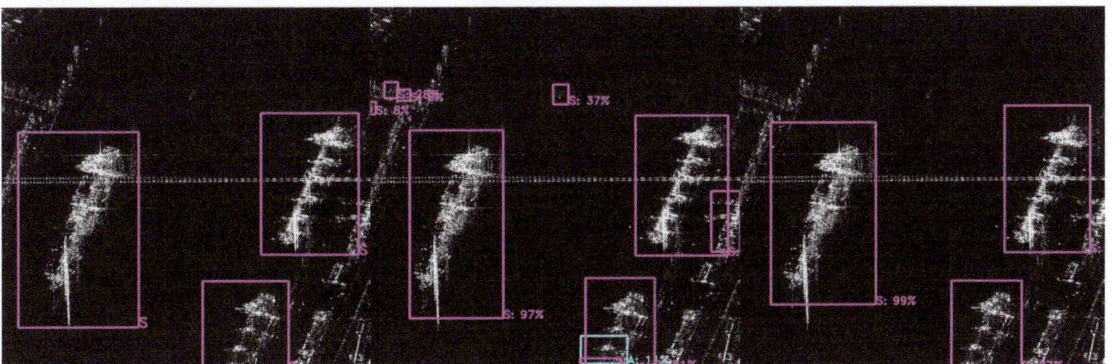

(c) Ship (S) class

Figure 9. Image triples are shown in which the left image is groundtruth, while the middle image is for baseline models (traditional two-stage detection models with real SAR images), and the right image is for our models. The groundtruth and predicted bounding boxes are plotted in blue color for A class, yellow color for E class, and pink color for S class. The numbers on the bounding boxes in the middle and right images denote the confidence score for each corresponding category. We visualize all detected bounding boxes after NMS and thresholding detector confidence at 0.05.

Table 2. Comparison of detection performance on our constructed dataset with TerraSAR-X and COSMO-SkyMed images. By incorporating region proposals from denoised SAR images within detection network, our model shows significant improvement in AP. The entries with the best APs for each object category are highlight in bold.

Backbone	+Despeckling	AP	Airplane (A)	Etcetera (E)	Ship (S)
ResNet-50	-	52.05	53.90	54.54	47.72
	preprocessing (Lee filter [22])	53.52	54.63	56.96	48.98
	preprocessing (PPB filter [25])	51.16	54.35	53.68	45.44
	within network (ours)	55.90	58.82	54.04	54.84
ResNet-101	-	54.29	54.65	59.80	48.43
	preprocessing (Lee filter [22])	56.19	58.04	60.59	49.95
	preprocessing (PPB filter [25])	52.96	53.16	58.17	47.54
	within network (ours)	60.81	65.03	61.67	55.72

3.5. Ablation Study

We conduct an ablation study for structurally verifying the proposed fusing region proposal strategy. We first compare the case without fusing itself after denoising on input noisy SAR image, which corresponds to the first experiment in Table 3. With the comparison to inputting only denoised SAR image as an input to detection network, we can identify whether the usage of real SAR image as another input of the detection network is important. This case shows the poorest detection performance and justifies the importance of fusing information from raw noisy SAR images. Secondly, for the choice of feature map after fusing, we perform experiments with feature map from denoised SAR image or feature map from real SAR image. As a result, keeping the feature map from the real SAR image as proposed is found to be much better.

Table 3. Ablation study across the input type of detection network and feature map forwarded to subsequent sub-network for classification and bounding box regression for each RoIs. The entries with the best APs for each object category are highlight in bold. The backbone is ResNet-50-FPN.

Input of DetNet.	Feature Map	AP	Airplane (A)	Etcetera (E)	Ship (S)
Denoised only	-	52.96	56.71	53.59	48.57
Real + Denoised	Denoised	53.96	57.16	51.17	53.54
Real + Denoised	Real (ours)	55.90	58.82	54.04	54.84

4. Discussion

Our proposed detection framework obviously achieves a better performance through combining a denoising network with an existing detection network; however, more parameters and the complex structure demand larger memory for model storage and higher computing cost. We report average inference times (measured in seconds/(patch image) on a Titan Xp GPU) for the purpose of time complexity analysis, as presented in Table 4. Compared with the existing two-stage object detection network like Faster RCNN [45] in the first row of Table 4, our detection framework further requires denoising time and time for fusing region proposals during inference. The denoising time makes up a large portion of the added running times, so the most promising way for reducing the average inference time would be adopting a relatively light denoising network.

Table 4. Comparison of running times for the time complexity analysis. We evaluated the running times on a patch image sized 800 × 800 with a Titan Xp GPU.

Models	Inference Time (sec/patch)
Faster RCNN [45]	0.3854
Faster RCNN + Ours	0.8190

5. Conclusions

In this study, we develop a novel object detection framework, where an unsupervised denoising network is combined with a two-stage detection network and two sets of region proposals extracted from a real noisy SAR image and a synthetically denoised SAR image are complementarily merged. The coupling structure of denoising network with detection network together intends to replace a cumbersome preprocessing step for denoising with our denoising network and at the same time, the integrated denoising network performs denoising to support the subsequent object detection. To remedy a potential risk due to fine information loss after denoising, we keep raw information from input SAR image within detection network while only utilize a set of region proposals inferred from the synthetically denosied SAR image. The extensive qualitative and quantitative experiments on our own datasets with TerraSAR-X and COSMO-SkyMed satellite images suggest that the proposed object detection framework involves the adaptive denoising for directly influencing detection performance. Our method shows significant improvements over several detection baselines on the datasets constructed from TerraSAR-X and COSMO-SkyMed satellite images.

Author Contributions: Conceptualization, all authors; methodology, S.S.; software, S.S.; validation, Y.K., I.H. and J.K.; formal analysis, J.K.; investigation, I.H.; resources, S.S.; data curation, S.S.; writing—original draft preparation, S.S.; writing—review and editing, S.S.; visualization, S.S. and Y.K.; supervision, S.K.; project administration, S.K. All authors have read and agreed to the published version of the manuscript.

Funding: This research received no external funding.

Institutional Review Board Statement: Not applicable.

Informed Consent Statement: Not applicable.

Data Availability Statement: Not applicable.

Acknowledgments: This research was supported by the Defense Challengeable Future Technology Program of Agency for Defense Development, Republic of Korea.

Conflicts of Interest: The authors declare no conflict of interest.

References

1. Chang, Y.L.; Anagaw, A.; Chang, L.; Wang, Y.C.; Hsiao, C.Y.; Lee, W.H. Ship detection based on YOLOv2 for SAR imagery. *Remote Sens.* **2019**, *11*, 786. [CrossRef]
2. Chen, P.; Li, Y.; Zhou, H.; Liu, B.; Liu, P. Detection of small ship objects using anchor boxes cluster and feature pyramid network model for SAR imagery. *J. Mar. Sci. Eng.* **2020**, *8*, 112. [CrossRef]
3. Kang, M.; Ji, K.; Leng, X.; Lin, Z. Contextual region-based convolutional neural network with multilayer fusion for SAR ship detection. *Remote Sens.* **2017**, *9*, 860. [CrossRef]
4. Lin, Z.; Ji, K.; Leng, X.; Kuang, G. Squeeze and excitation rank faster R-CNN for ship detection in SAR images. *IEEE Geosci. Remote Sens. Lett.* **2018**, *16*, 751–755. [CrossRef]
5. Zhai, L.; Li, Y.; Su, Y. Inshore ship detection via saliency and context information in high-resolution SAR images. *IEEE Geosci. Remote Sens. Lett.* **2016**, *13*, 1870–1874. [CrossRef]
6. He, C.; Tu, M.; Xiong, D.; Tu, F.; Liao, M. A component-based multi-layer parallel network for airplane detection in SAR imagery. *Remote Sens.* **2018**, *10*, 1016. [CrossRef]
7. Zhang, L.; Li, C.; Zhao, L.; Xiong, B.; Quan, S.; Kuang, G. A cascaded three-look network for aircraft detection in SAR images. *Remote Sens. Lett.* **2020**, *11*, 57–65. [CrossRef]

8. Diao, W.; Dou, F.; Fu, K.; Sun, X. Aircraft detection in sar images using saliency based location regression network. In Proceedings of the IGARSS 2018 IEEE International Geoscience and Remote Sensing Symposium, Valencia, Spain, 22–27 July 2018; pp. 2334–2337.
9. Zhao, Y.; Zhao, L.; Li, C.; Kuang, G. Pyramid attention dilated network for aircraft detection in SAR images. *IEEE Geosci. Remote Sens. Lett.* **2021**, *18*, 662–666. [CrossRef]
10. Saha, S.; Bovolo, F.; Bruzzone, L. Destroyed-buildings detection from VHR SAR images using deep features. In *Image and Signal Processing for Remote Sensing XXIV*; International Society for Optics and Photonics: Bellingham, WA, USA, 2018; Volume 10789, p. 107890Z.
11. Bao, S.; Meng, J.; Sun, L.; Liu, Y. Detection of ocean internal waves based on Faster R-CNN in SAR images. *J. Oceanol. Limnol.* **2020**, *38*, 55–63. [CrossRef]
12. Huang, D.; Du, Y.; He, Q.; Song, W.; Liotta, A. DeepEddy: A simple deep architecture for mesoscale oceanic eddy detection in SAR images. In Proceedings of the 2017 IEEE 14th International Conference on Networking, Sensing and Control (ICNSC), Calabria, Italy, 16–18 May 2017; pp. 673–678.
13. Bianchi, F.M.; Espeseth, M.M.; Borch, N. Large-scale detection and categorization of oil spills from SAR images with deep learning. *Remote Sens.* **2020**, *12*, 2260. [CrossRef]
14. Waldeland, A.U.; Reksten, J.H.; Salberg, A.B. Avalanche detection in sar images using deep learning. In Proceedings of the IGARSS 2018 IEEE International Geoscience and Remote Sensing Symposium, Valencia, Spain, 22–27 July 2018; pp. 2386–2389.
15. Rotter, P.; Muron, W. Automatic Detection of Subsidence Troughs in SAR Interferograms Based on Convolutional Neural Networks. *IEEE Geosci. Remote Sens. Lett.* **2020**, *18*, 82–86. [CrossRef]
16. Sun, X.; Wang, Z.; Sun, Y.; Diao, W.; Zhang, Y.; Fu, K. AIR-SARShip-1.0: High-resolution SAR Ship Detection Dataset. *J. Radars* 2019, *8*, 852–862. (In English)
17. Wang, Y.; Wang, C.; Zhang, H.; Dong, Y.; Wei, S. A SAR Dataset of Ship Detection for Deep Learning under Complex Backgrounds. *Remote Sens.* **2019**, *11*, 765. [CrossRef]
18. Li, J.; Qu, C.; Shao, J. Ship detection in SAR images based on an improved faster R-CNN. In Proceedings of the 2017 SAR in Big Data Era: Models, Methods and Applications (BIGSARDATA), Beijing, China, 13–14 November 2017; pp. 1–6.
19. Wei, S.; Zeng, X.; Qu, Q.; Wang, M.; Su, H.; Shi, J. HRSID: A high-resolution SAR images dataset for ship detection and instance segmentation. *IEEE Access* **2020**, *8*, 120234–120254. [CrossRef]
20. Bai, Y.C.; Zhang, S.; Chen, M.; Pu, Y.F.; Zhou, J.L. A fractional total variational CNN approach for SAR image despeckling. In *International Conference on Intelligent Computing*; Springer: Berlin/Heidelberg, Germany, 2018; pp. 431–442.
21. Parrilli, S.; Poderico, M.; Angelino, C.V.; Verdoliva, L. A nonlocal SAR image denoising algorithm based on LLMMSE wavelet shrinkage. *IEEE Trans. Geosci. Remote Sens.* **2011**, *50*, 606–616. [CrossRef]
22. Lee, J.S. Speckle analysis and smoothing of synthetic aperture radar images. *Comput. Graph. Image Process.* **1981**, *17*, 24–32. [CrossRef]
23. Kuan, D.T.; Sawchuk, A.A.; Strand, T.C.; Chavel, P. Adaptive noise smoothing filter for images with signal-dependent noise. *IEEE Trans. Pattern Anal. Mach. Intell.* **1985**, *PAMI-7*, 165–177. [CrossRef] [PubMed]
24. Frost, V.S.; Stiles, J.A.; Shanmugan, K.S.; Holtzman, J.C. A model for radar images and its application to adaptive digital filtering of multiplicative noise. *IEEE Trans. Pattern Anal. Mach. Intell.* **1982**, *PAMI-4*, 157–166. [CrossRef]
25. Deledalle, C.A.; Denis, L.; Tupin, F. Iterative weighted maximum likelihood denoising with probabilistic patch-based weights. *IEEE Trans. Image Process.* **2009**, *18*, 2661–2672. [CrossRef]
26. Guo, H.; Wu, D.; An, J. Discrimination of oil slicks and lookalikes in polarimetric SAR images using CNN. *Sensors* **2017**, *17*, 1837. [CrossRef] [PubMed]
27. Xu, Q.; Li, W.; Xu, Z.; Zheng, J. Noisy SAR image classification based on fusion filtering and deep learning. In Proceedings of the 2017 3rd IEEE International Conference on Computer and Communications (ICCC), Chengdu, China, 13–16 December 2017; pp. 1928–1932.
28. Zhang, T.; Zhang, X. High-speed ship detection in SAR images based on a grid convolutional neural network. *Remote Sens.* **2019**, *11*, 1206. [CrossRef]
29. Hong, S.J.; Baek, W.K.; Jung, H.S. Ship Detection from X-Band SAR Images Using M2Det Deep Learning Model. *Appl. Sci.* **2020**, *10*, 7751. [CrossRef]
30. Zhao, J.; Zhang, Z.; Yu, W.; Truong, T.K. A cascade coupled convolutional neural network guided visual attention method for ship detection from SAR images. *IEEE Access* **2018**, *6*, 50693–50708. [CrossRef]
31. Versaci, M.; Morabito, F.C.; Angiulli, G. Adaptive image contrast enhancement by computing distances into a 4-dimensional fuzzy unit hypercube. *IEEE Access* **2017**, *5*, 26922–26931. [CrossRef]
32. Orujov, F.; Maskeliūnas, R.; Damaševičius, R.; Wei, W. Fuzzy based image edge detection algorithm for blood vessel detection in retinal images. *Appl. Soft Comput.* **2020**, *94*, 106452. [CrossRef]
33. Wang, J.; Zheng, T.; Lei, P.; Bai, X. Ground target classification in noisy SAR images using convolutional neural networks. *IEEE J. Sel. Top. Appl. Earth Obs. Remote Sens.* **2018**, *11*, 4180–4192. [CrossRef]
34. German Aerospace Center. Available online: https://www.dlr.de/EN/Home/home_node.html (accessed on 4 May 2021).
35. Italian Space Agency. Available online: https://www.asi.it/en/#divFooter (accessed on 4 May 2021).
36. Laine, S.; Karras, T.; Lehtinen, J.; Aila, T. High-Quality Self-Supervised Deep Image Denoising. *Adv. Neural Inf. Process. Syst.* **2019**, *32*, 6970–6980.

37. Molini, A.B.; Valsesia, D.; Fracastoro, G.; Magli, E. Speckle2Void: Deep Self-Supervised SAR Despeckling with Blind-Spot Convolutional Neural Networks. *IEEE Trans. Geosci. Remote Sens.* **2021**. [CrossRef]
38. Khademi, W.; Rao, S.; Minnerath, C.; Hagen, G.; Ventura, J. Self-supervised poisson-gaussian denoising. In Proceedings of the IEEE/CVF Winter Conference on Applications of Computer Vision, Waikoloa, HI, USA, 5–9 January 2021; pp. 2131–2139.
39. Chen, K.; Wang, J.; Pang, J.; Cao, Y.; Xiong, Y.; Li, X.; Sun, S.; Feng, W.; Liu, Z.; Xu, J.; et al. MMDetection: Open MMLab Detection Toolbox and Benchmark. *arXiv* **2019**, arXiv:1906.07155.
40. Paszke, A.; Gross, S.; Massa, F.; Lerer, A.; Bradbury, J.; Chanan, G.; Killeen, T.; Lin, Z.; Gimelshein, N.; Antiga, L.; et al. PyTorch: An Imperative Style, High-Performance Deep Learning Library. In Proceedings of the Advances in Neural Information Processing Systems 32, Vancouver, BC, Canada, 8–14 December 2019; pp. 8024–8035.
41. Robbins, H.; Monro, S. A stochastic approximation method. *Ann. Math. Stat.* **1951**, *22*, 400–407. [CrossRef]
42. Sutskever, I.; Martens, J.; Dahl, G.; Hinton, G. On the importance of initialization and momentum in deep learning. In Proceedings of the International Conference on Machine Learning, Atlanta, GA, USA, 16–21 June 2013; pp. 1139–1147.
43. He, K.; Zhang, X.; Ren, S.; Sun, J. Deep residual learning for image recognition. In Proceedings of the IEEE Conference on Computer Vision and Pattern Recognition, Las Vegas, NV, USA, 27–30 June 2016; pp. 770–778.
44. Lin, T.Y.; Dollár, P.; Girshick, R.; He, K.; Hariharan, B.; Belongie, S. Feature pyramid networks for object detection. In Proceedings of the IEEE Conference on Computer Vision and Pattern Recognition, Honolulu, HI, USA, 21–26 July 2017; pp. 2117–2125.
45. Ren, S.; He, K.; Girshick, R.; Sun, J. Faster r-cnn: Towards real-time object detection with region proposal networks. *IEEE Trans. Pattern Anal. Mach. Intell.* **2016**, *39*, 1137–1149. [CrossRef] [PubMed]

applied sciences

Article

Rotational-Shearing-Interferometer Response for a Star-Planet System without Star Cancellation

Beethoven Bravo-Medina [1,*], Marija Strojnik [2], Azael Mora-Nuñez [1] and Héctor Santiago-Hernández [1]

[1] Departamento de Electrónica, Universidad de Guadalajara, Av. Revolución 1500, C.P. 44840 Guadalajara, Jalisco, Mexico; dejesus.mora@academicos.udg.mx (A.M.-N.); hector.santiagoh@academicos.udg.mx (H.S.-H.)
[2] Centro de InCPvestigaciones en Óptica, Apdo. Postal 1-948, C.P. 37000 León, Guanajuato, Mexico; mstrojnik@gmail.com
* Correspondence: beethovenbm@gmail.com

Abstract: The Rotational Shearing Interforometer has been proposed for direct detection of extra-solar planets. This interferometer cancels the star radiation using destructive interference. However, the resulting signal is too small (few photons/s for each m^2). We propose a novel method to enhance the signal magnitude by means of the star–planet interference when the star radiation is not cancelled. We use interferograms computationally simulated to confirm the viability of the technique.

Keywords: interferometry; remote sensing; computational simulation

Citation: Bravo-Medina, B.; Strojnik, M.; Mora-Nuñez, A.; Santiago-Hernández, H. Rotational-Shearing-Interferometer Response for a Star-Planet System without Star Cancellation. *Appl. Sci.* **2021**, *11*, 3322. https://doi.org/10.3390/app11083322

Academic Editor: Yang Dam Eo

Received: 10 March 2021
Accepted: 1 April 2021
Published: 7 April 2021

1. Introduction

In the last twenty years, the interest in exoplanet detection has been increased. More than 4000 planets have been discovered [1], The main interest of the scientific community are the earthlike planets [2].

The overwhelming majority of discovering planet was discovered by indirect techniques such as transit light curves [3], radial velocity [4], and gravitational microlensing [5]. These techniques sense the effect of the planet over the star radiation. They measure variations in the star radiation and determine the planet presence using statistical techniques. However, these variations may be produced by unknown processes in the star and not by the presence of a planet. Additionally, the measurements obtained by the microlensing technique are not repeatable because they require the alignment of two stars and the planet. Moreover, the time necessary to perform a measurement employing indirect techniques may last from days to years because the planet must complete at least one orbit.

The direct detection of an exoplanet will confirm the currently available evidence for its existence, with shorter observation periods and incorporating the repeatability ofthe measurement.

The principal challenges in the direct detection of exoplanets are the image resolution and the signal-to-noise ratio. The image resolution refers to the minimum angular distance between two discernible sources. This angular distance is conditioned by the diameter of the primary mirror of the telescope. The signal-to-noise ratio may be defined as the quotient between the planet radiance and the star radiance. At visible wavelengths, the planet radiance arises primarily from the radiation reflected from its parent star. The amount of reflected radiation depends on the planet's albedo, its radius, and its distance from the star. This ratio for a Jupiter-like planet and a Sun-like star is at least 10^{-10} [6]. In the infrared (IR) spectral region, the planet radiance consists primarily of the planet thermally emitted radiation. Additionally, the radiance for a Sun-like star is lower in the IR than in the visible region. Under these conditions, the radiation signal-to-noise ratio increases up to 10^{-5} [7]. However, this is still a very low signal-to-noise ratio. In order to improve the radiation ratio, a coronagraph and interferometric techniques are often implemented [8,9].

The coronagraph technique consists of occulting the star using a mechanical aperture. This technique usually implements spatial transmission filters in the focal plane to remove the diffraction rings due to the hard stop edge [10,11]. However, due to the limitations of the spatial resolution, this technique is applied primarily to the potential planets with large orbits.

Most of interferometric techniques attenuate the star radiation by means of destructive interference. Consequently, these interferometers are called Nulling Interferometers. They interfere with the wavefronts with a delayed version of them. When the delay between the interferometer arms is $\lambda/2$, the star radiation is canceled, with only the planet radiation remaining [12–14].

Previously, we proposed a Rotational Shearing Interferometer (RSI) for planet detection [15–18]. This interferometer interferes with the wavefronts with a rotated and delayed version of them. Thus, this interferometer may cancel the star radiation in a similar way to the nulling interferometers. Additionally, we may discriminate against false-positive results rotating the wavefront. Unfortunately, when the star is canceled, the remaining amount of radiation is too small. In order to improve the signal magnitude, we propose using the RSI without the total-cancellation of the star radiation. When the star-radiation is not canceled, the fringe visibility is decreased, but the signal magnitude is further increased.

In this work, we describe the response of the RSI to a star–planet system radiation in Section 2. In Section 3, we present computational simulations to validate the viability of the method. Finally, we present the conclusions.

2. Theory

The closest star to Earth is Proxima Centauri, its distance to the Earth is 1.295 parsecs [19]. At this distance, the optical radiation from a massive light source like a star or a planet may be considered coherent. This is because these conditions satisfy the Van Cittert–Zernike theorem [20]. Then, the radiation from any planetary-system outside the solar system may interfere between them. Additionally, the star and the planet may be considered as point sources [21]. Figure 1 shows a diagram of the star–planet system viewed for the observer. The alignment of the planet and the star is characterized by means of their elevation angle (θ) and their azimuth angle (φ). The wavefronts from a star or a planet in its periphery may be modeled as planes with uniform intensity due to the long distance from the observer.

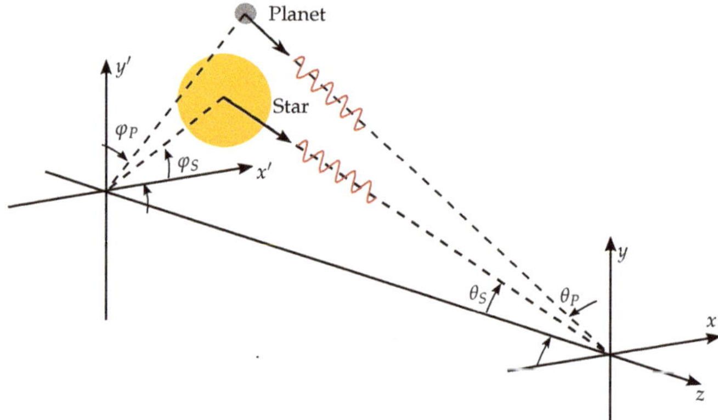

Figure 1. Star–planet system viewed for the observer. The star and the planet may be modeled as point sources with an elevation (θ) and an azimuth (φ) angle with respect to the optical axis.

The Rotational Shearing Interferometer (RSI) was proposed for extra-solar planet detection because it is insensitive to rotational-symmetrically wavefronts. The RSI makes incident wavefront interfere with a rotated version of them. When a solitary star is aligned on the optical axis of the interferometer, its wavefront is rotational-symmetrically, and the RSI does not produce a fringe pattern. Instead, when the wavefronts from the star and the planet are incident on the RSI, its interference produces a fringe pattern. The RSI consists of a Mach–Zehnder interferometer with a Dove prism in each arm as shown in Figure 2. When the Dove prism rotates around the optical axis, the propagated wavefront rotates double of the Dove-prism rotation angle. One of the Dove prism is rotated in order to generate the wavefront rotation. The other Dove prism remains static for compensation purposes. We use a mirror array as an Optical Path Modulator (OPM). This array is formed by the mirrors M_4, M_5, and M_6. The mirrors M_5 and M_6 are collocated over a displacement platform to control the elongation of the optical Path. We add an additional mirror array in the other arm for compensation purposes. When a beam insides on the first beam splitter, it is divided in two. The first one is propagated trough the prism DP_1, and it is rotated by $\Delta\phi$ with respect to the other beam. The second beam is propagated through the OPM to adjust the OPD. Finally, both beams interfere in the observation plane (OP).

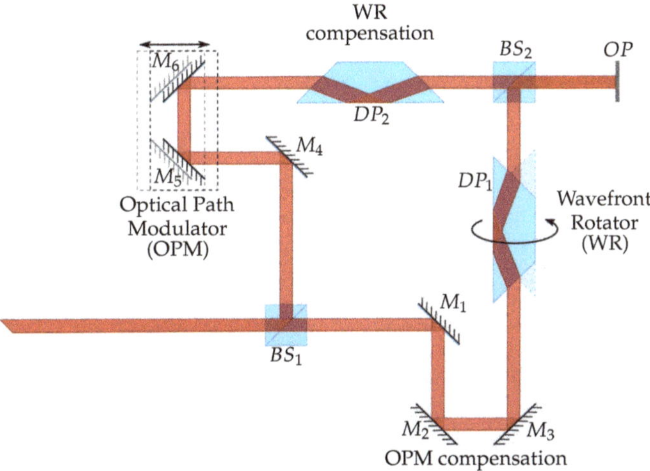

Figure 2. The Rotational Shearing Interferometer consists of a modified Mach–Zehnder interferometer, with a Dove prism as Wavefront Rotator (WR), a mirror array as an Optical Path Modulator (OPM), and its respective compensation components in the opposite arm.

When a beam coming from a star–planet system is incident on the RSI entrance, four wavefronts interfere between them on the observation plane. Two of these wavefronts correspond to the planet and the other two correspond to the star. The resulting interference may be divided into three terms: the first one for the interference between the star wavefronts (M_{SS}), the second one for the interference between the planet wavefronts (M_{PP}), and an additional term for the interference between the planet and the star wavefronts (M_{SP}). Then, the incidance (M) in the interference plane may be modeled as the sum of the terms:

$$M = M_{SS} + M_{PP} + M_{SP}. \tag{1}$$

Specifically, these terms may be written in terms of the incidance of the star (M_P), incidance of the planet (M_P) and phase (Φ_{ij}) of each wavefront. The subscripts $i = S, P$ and $j = 1, 2$, represent the wavefronts, and the interferometer arms, respectively:

$$M_{SS} = M_S[2 + 2\cos(\Phi_{S2} - \Phi_{S1})], \tag{2}$$

$$M_{PP} = M_P[2 + 2\cos(\Phi_{P2} - \Phi_{P1})], \tag{3}$$

$$M_{SP} = \sqrt{M_S M_P}[\cos(\Phi_{S1} - \Phi_{P1}) + \cos(\Phi_{S1} - \Phi_{P2}) + \cos(\Phi_{S2} - \Phi_{P1}) + \cos(\Phi_{S2} - \Phi_{P2})]. \quad (4)$$

Furthermore, the phase term ($\Phi_{ij}(\rho, \theta, \varphi)$) in cylindrical coordinates for each wavefront is modeled by:

$$\Phi_{S1} = \omega t + \frac{2\pi}{\lambda}\rho \sin\theta_S \cos(\varphi - \varphi_S) + L_1 \cos\theta_S, \quad (5a)$$

$$\Phi_{S2} = \omega t + \frac{2\pi}{\lambda}\rho \sin\theta_S \cos(\varphi - \varphi_S + 2\Delta\varphi) + L_2 \cos\theta_S, \quad (5b)$$

$$\Phi_{P1} = \omega t + \frac{2\pi}{\lambda}\rho \sin\theta_P \cos(\varphi - \varphi_S) + L_1 \cos\theta_P, \quad (5c)$$

$$\Phi_{P2} = \omega t + \frac{2\pi}{\lambda}\rho \sin\theta_P \cos(\varphi - \varphi_S + 2\Delta\varphi) + L_2 \cos\theta_P. \quad (5d)$$

We use L_i to denote the optical path length of each interferometer arm. The terms θ_S and φ_S are used to indicate the elevation and the azimuthal angles between the star and the optical axis, respectively. In similar way, θ_P and φ_P are the angles between the planet and the optical axis.

2.1. Special Cases

In order to simplify the analysis of these equations, we consider three special cases. In the first case, the star does not have a planet around it ($M_P = 0$), and the optical-path-difference (OPD) of the interferometer ($L_2 - L_1$) is equal to $\lambda/2$. This case is illustrated in Figure 3. The second case occurs when the optical axis aligned on the star ($\theta_S = 0$) and the OPD is $\lambda/2$; in this case, the star wavefronts are canceled. Finally, we consider the case when the star is perfectly aligned with the star, but the OPD is different to $\lambda/2$ (the star wavefront is not canceled). Figure 4 illustrates the star and planet alignment for the second case and the third case.

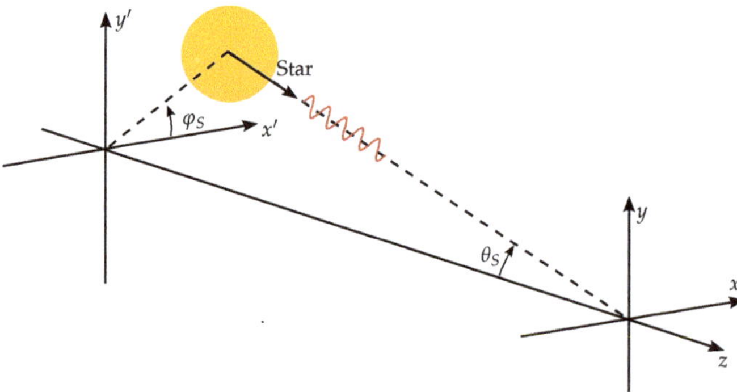

Figure 3. Furthermore, the phase term ($\Phi_{ij}(\rho, \theta, \varphi)$) in cylindrical coordinates for each wavefront is modeled by:

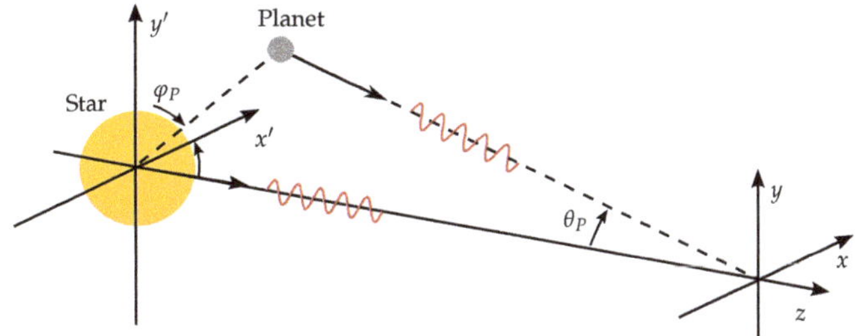

Figure 4. Star–planet system viewed for the interferometer when the star is aligned with the interferometer optical axis.

2.1.1. First Case, the Solitary Star

When the star does not have a planet as a companion, the interference equation is reduced to the term M_{SS}, described by the next equation:

$$M = M_{SS} = M_S\left(2 + 2\cos\left\{\frac{2\pi}{\lambda}[2\rho\sin\theta_S\sin\Delta\varphi\sin(\varphi - \varphi_S - \Delta\varphi) + OPD\cos\theta_S]\right\}\right). \tag{6}$$

Equation (6) represents a fringe pattern. The spatial frequency of the fringes is given by $2\theta_S\sin\Delta\varphi$. This equation denotes the dependency of the fringe density with the elevation angle of the star. The orientation of the fringe-pattern is given by $\varphi_S - \Delta\varphi$. Note that the azimuth angle of the star determines the fringe direction. Because of the dependence between the fringe pattern and the star alignment, we may use this fringe pattern to align the optical axis with the star. When the star is perfectly aligned with the optical axis, the resultant pattern consists of a uniform incidance over the entire observation plane. If we change the interferometer OPD, the incidance level varies accordingly with:

$$M = M_S\left[2 + 2\cos\left(\frac{2\pi}{\lambda}OPD\right)\right]. \tag{7}$$

If we adjust the OPD to $\lambda/2$, the star incidance is canceled at the observation plane. Under these conditions, it is possible to detect a planet if it is present around the planet.

2.1.2. Second Case, Star on Axis and OPD Equal to $\lambda/2$

When the star is aligned with the optical axis and it is orbited by a planet, the RSI receives the star and the planet wavefronts simultaneously. If additionally, we adjust the OPD to $\lambda/2$, the star wavefront is canceled by destructive interference. This case is analyzed in the most of interferometric methods to detect extra-solar planets. The incidance equation is considerably reduced because the terms M_{SS} and M_{SP} are canceled; additionally, θ_P is small enough to use the paraxial approximation. The resulting equation may be rewritten as:

$$M = M_P\left(2 + 2\cos\left[\frac{2\pi}{\lambda}2\rho\theta_P\sin\Delta\varphi\sin(\varphi - \varphi_P - \Delta\varphi) + \pi\right]\right). \tag{8}$$

Equation (8) represents a fringe pattern with spatial frequency equal to $2\theta_P\sin\Delta\varphi$ and orientation equal to $\varphi_P - \Delta\varphi$. Both expressions are dependent on the rotation angle, $\Delta\varphi$. This demonstrates that the frequency and orientation on the fringes may be controlled by the operator as it is shown in laboratory implementations [22]. The maximum fringe density is reached when the angle between the Dove prisms is 90°. The minimum fringe separation is $\frac{\lambda}{2\theta_p}$ (about 2 m for a Jupiter-like planet at 10 parsecs from the Earth, observed at 10 µm).

In this case, the fringe visibility is only limited by the coherence function of the incident beam. Unfortunately, the incidance is too low because the planet incidance M_P is just a few photons/s per m^2 [15].

2.1.3. Third Case, Star on Axis and OPD $\neq \lambda/2$

In order to increase the signal incidance, we propose using the RSI without a total star cancellation. In these conditions, the fringe visibility is reduced, but the signal amplitude is enhanced considerably. If we consider the star on axis ($\theta_S = 0$) and ignore some phase terms, we may simplify Equations (2)–(4) to the next equations:

$$M_{SS} = M_S\left\{2 + 2\cos\left[\frac{2\pi}{\lambda}OPD\right]\right\}, \tag{9}$$

$$M_{PP} = M_P\left(2 + 2\cos\left\{\frac{2\pi}{\lambda}[2\rho\sin\theta_P\sin\Delta\varphi\sin(\varphi - \varphi_P - \Delta\varphi) + OPD\cos\theta_P]\right\}\right), \tag{10}$$

$$M_{SP} = 4\sqrt{M_S M_P}\cos\left(\frac{2\pi}{\lambda}\frac{OPD}{2}\right)\left(\cos\left\{\frac{2\pi}{\lambda}[\rho\sin\theta_P\sin(\varphi - \varphi_P) + L_1\cos\theta_P]\right\}\right. \tag{11}$$
$$\left. + \cos\left\{\frac{2\pi}{\lambda}[\rho\sin\theta_P\sin(\varphi - \varphi_P + \Delta\varphi) + L_2\cos\theta_P]\right\}\right).$$

Note that Equation (9) is equal to Equations (7) and (10) is equal to Equation (8). Accordingly, these terms produce a background incidance and a fringe pattern. Furthermore, the interference between the planet and the star, represented by Equation (11), produces two superposed fringe patterns. The first one oriented to φ_P and the second one oriented to $\varphi_p - \Delta\varphi$. Their magnitude is modulated by the cosine of $\pi OPD/\lambda$. Their spatial frequency is $\rho\sin\theta_P$, and their fringe separation is $\lambda\theta_p$ (about 4 m for a Jupiter-like planet at 10 parsecs from the Earth, observed at 10 µm). The fringe visibility is reduced for the background incidance, which is increased; accordingly, the OPD moves away $\lambda/2$. However, the amplitude of the fringe patterns generated by M_{SP} is increased too. The decrease of the fringe visibility could persuade the researcher of this way. Notwithstanding, however as long as the detector is not saturated, the signal may be retrieved by image processing. In this way, we may amplify the signal several times until the detector saturates. The maximum signal amplification depends on the amount of bits of the detector.

3. Computational Simulation

In order to verify the advantages of the proposed technique, we perform a computational simulation of the RSI and its response to a star–planet system. We use an exact ray trace over the RSI to determine the wavefront modification. The wavefront was simulated using three rays whose sources are located over a plain. The rays are propagated in parallel to the propagation vector of the wavefront. On each surface, a new ray set is calculated according to the reflection or refraction laws as appropriate. When the rays insides over the observation plane, their optical path length is calculated. Using this information, we determine the wavefront transformation after they have been propagated by the RSI. The process is repeated for each wavefront and each interferometer arm. The incidance at each point of the observation plane is calculated using the incidance and phase of each incident wavefront. Finally, the resultant interferogram is determined by mapping the resultant incidance with a grayscale value. This computational simulation technique was explained with more detail in [23].

We simulate a star–planet system where the angular distance between the star and the planet is 0.5 arcsec, and the star radiation is 10^5 times the planet radiation at 1 µm. The star is perfectly aligned with the optical axis. The azimuth angle of the planet is 0. The observation plane dimensions are 1 m x 1 m. We use these characteristics to probe the improvement in the planet detection performed by the proposed technique. However, equivalent advantages may be achieved with any star–planet system.

Figure 5 shows six interferograms obtained by computational simulation. The interferograms were generated by adjusting the OPD to $\lambda/2$ and changing the rotation angle from 0° to 180° with the star aligned with the optical axis. These interferograms are composed of straight fringes for which density and orientation change when the interferometer rotation-angle is changed, according to Equation (8). The fringes are produced by the interference of the planet wavefront with itself; this confirms the planet presence. The rotation of the fringe allows for discarding several false-positives by alignment errors. The image grayscale range is adjusted to coincide its saturation level with the maximum incidance, produced by M_{PP} term.

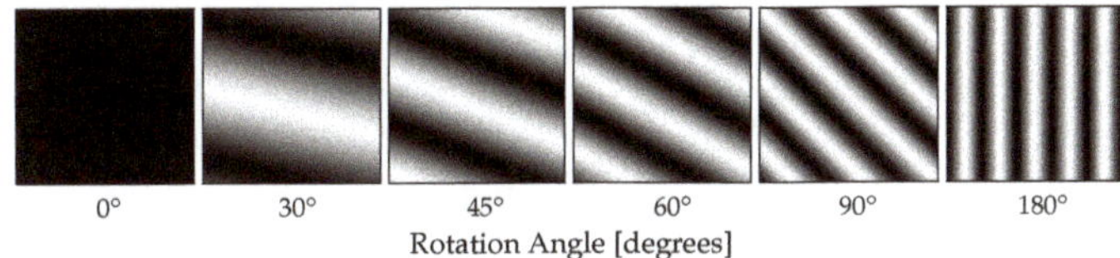

Figure 5. Simulated interferograms obtained as the response of a RSI to a star–planet system wavefronts when the star is aligned with the optical axis and the OPD is $\lambda/2$. The rotation angle of the RSI is indicated below each interferogram. The grayscale range is adjusted to coincide its saturation level with the maxim of M_{PP} term.

Figure 6 shows interferograms generated adjusting the OPD to $\lambda/2 + 1$ nm with the star aligned to the optical axis. These interferograms are composed by two straight fringes superposed. The fringes density and orientation changes with the interferometer rotation-angle as predicted in Equation (8). These fringes correspond to the interference between the star and the planet. The fringes produced by the interference of the planet with itself are present; however, they are eclipsed by the brightness of star–planet fringes. The first image shows the background incidance produced by the interference of the star wavefront with its rotated version. The image grayscale is adjusted to coincide with the maximum incidance, produced by the M_{SP} term. In this case, we obtain a signal gain of 4 with respect to the previous case. The gain is calculated as the difference between the minimum and the maximum incidance level of the resultant interferogram compared to the difference between the minimum and the maximum of the M_{PP} term.

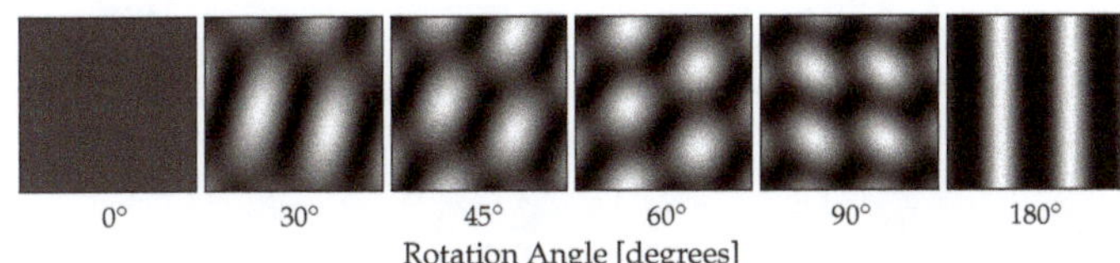

Figure 6. Response of a RSI when its optical axis is aligned with the star and the OPD is $\lambda/2 + 1$nm. The RSI rotation-angle is indicated below each interferogram. The image grayscale is adjusted to coincide its saturation level with the maximum of the M_{SP} term.

The increment in the fringe incidance in accordance with the OPD variation is showed by Figure 7. It shows interferograms with different grayscale ranges: in the first row, the image saturation-level is 8 times the M_{PP} maximum, in the second row 64 times, in the third row 128 times, and 256 times in the fourth row. The OPD is changed in each column: in the first column, the OPD is $\lambda/2$, in the second column, the OPD is $\lambda/2 + 2$ nm, in the third column, the OPD is $\lambda/2 + 5$ nm, in the fourth column, the OPD is $\lambda/2 + 10$ nm, and, finally, in the last column, the OPD is $\lambda/2 + 15$ nm. We may observe that the fringe visibility decreases accordingly the OPD moves away $\lambda/2$. However, the signal magnitude is increased and the pattern is still visible to the naked eye. When the OPD is $\lambda/2 + 15$ nm, the gain is 60 and the fringe visibility is 13%.

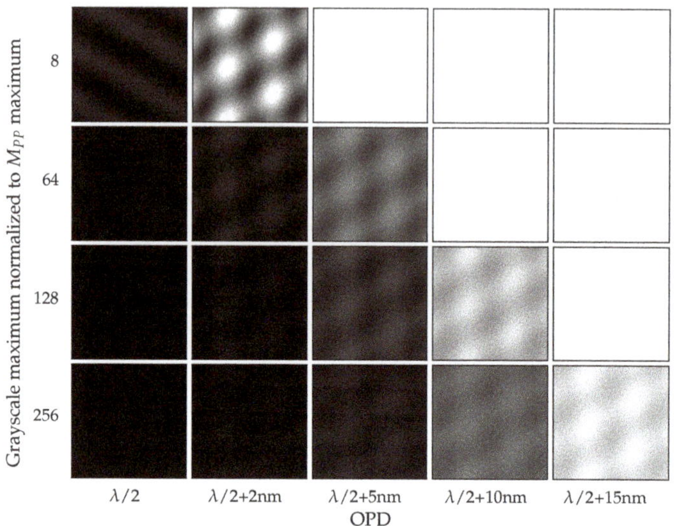

Figure 7. Comparison of the incidance level obtained for different values of OPD. We observe that the incidance level is incremented when the OPD moves away $\lambda/2$, and the fringe pattern is visible if we adjust adequately the grayscale range. The interferograms were obtained using a rotation angle of 60°. In each row, the grayscale maximum is adjusted to coincide with the value indicated on the left side of the row. The OPD used to simulate the interferograms of each column is indicated below it.

Figure 8 shows the fringe visibility and the gain versus the OPD. We may observe that the gain is increased almost linearly. In contrast, the fringe visibility decays rapidly. This behavior may confuse and erroneously discourage this technique because the visibility had a small value. However, the signal may be easily distinguishable, and the gain improvement is substantial as shown in Figure 7. Additionally, we may observe that the fringe visibility reduction is slow after 20 nm away $\lambda/2$, and the signal gain continues to increase at the same rate.

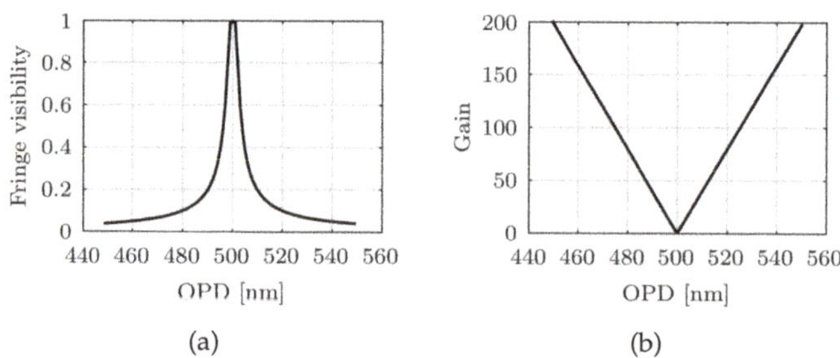

Figure 8. Fringe visibility (a) and signal gain (b) versus the OPD. When the OPD is moved away to detune from $\lambda/2$, the fringe visibility decreases, and the signal gain increases linearly at a rate of 4/nm.

4. Conclusions and Future Work

We proposed an improvement to the capacity of the RSI for extra-solar planets detection. This technique consists of the non-total cancellation of the star radiation in order to improve the signal magnitude. The signal increment is due to the interference of the planet wavefront with the star wavefront. We use computational simulations to demonstrate that the signal magnitude may be amplified more than 60 times and the signal may be still detected with the naked eye. The maximum signal gain is limited by the saturation level of the detector.

The search for direct detection of extra-solar planets is a long-term project. The next challenge consists of validating this technique by laboratory experiments.

Author Contributions: Conceptualization, B.B.-M. and M.S.; methodology, B.B.-M.; software, A.M.-N.; formal analysis, H.S.-H.; All authors have read and agreed to the published version of the manuscript.

Funding: Air Force Office of Scientific Research (FA9550-18-1-0454).

Institutional Review Board Statement: Not applicable.

Informed Consent Statement: Not applicable.

Data Availability Statement: The data presented in this study are available in the article.

Conflicts of Interest: The authors declare no conflict of interest.

Abbreviations

The following abbreviations are used in this manuscript:

RSI	Rotational Shearing Interferometer
OPD	Optical Path Difference
WR	Wavefront Rotator
OPM	Optical Path Modulator
M	Mirror
DP	Dove Prism
BS	Beam Splitter
OP	Observation Plane

References

1. Alei, E.; Claudi, R.; Bignamini, A.; Molinaro, M. Exo-MerCat: A merged exoplanet catalog with Virtual Observatory connection. *Astron. Comput.* **2020**, *31*, 100370. [CrossRef]
2. Checlair, J.H.; Villanueva, G.L.; Hayworth, B.P.C.; Olson, S.L.; Komacek, T.D.; Robinson, T.D.; Popovic, P.; Yang, H.; Abbot, D.S. Probing the capability of future direct imaging missions to spectrally constrain the frequency of Earth-like planets. *Astron. J.* **2021**, *161*, 150. [CrossRef]
3. Araújo, A.; Valio, A. Kepler-411 Differential Rotation from Three Transiting Planets. *Astrophys. J.* **2021**, *907*, L5. [CrossRef]
4. Kunder, A.; Kordopatis, G.; Steinmetz, M.; Zwitter, T.; McMillan, P.J.; Casagrande, L.; Enke, H.; Wojno, J.; Valentini, M.; Chiappini, C.; et al. The Radial Velocity Experiment (Rave): Fifth Data Release. *Astron. J.* **2017**, *153*, 75. [CrossRef]
5. Gaudi, B.S. Microlensing Surveys for Exoplanets. *Annu. Rev. Astron. Astrophys.* **2012**, *50*, 411–453. [CrossRef]
6. Strojnik, M.; Bravo-Medina, B. Simulation of Extrasolar Planet Detection with Rotationally Shearing Interferometer at 10 µm. *Multidiscip. Digit. Publ. Inst. Proc.* **2019**, *27*, 44. [CrossRef]
7. Vasquez-Jaccaud, C.; Strojnik, M.; Paez, G. Effects of a star as an extended body in extra-solar planet search. *J. Mod. Opt.* **2010**, *57*, 1808–1814. [CrossRef]
8. Beuzit, J.L.; Mouillet, D.; Oppenheimer, B.R.; Monnier, J.D. Direct Detection of Exoplanets. Invited review at the "Protostars and Planets V" Conference. In Proceedings of the PPV Conference, Hilton Waikoloa Village, HI, USA, 24–28 October 2005; Reipurth, B., Jewitt, D., Eds.; 2005, in press.
9. Fischer, D.A.; Howard, A.W.; Laughlin, G.P.; Macintosh, B.; Mahadevan, S.; Sahlmann, J.; Yee, J.C. Exoplanet detection techniques. *arXiv* **2015**, arXiv:1505.06869.
10. Mugnier, L.M.; Cornia, A.; Sauvage, J.F.; Rousset, G.; Fusco, T.; Védrenne, N. Optimal method for exoplanet detection by angular differential imaging. *J. Opt. Soc. Am. A* **2009**, *26*, 1326–1334. [CrossRef] [PubMed]
11. Cagigas, M.A.; Valle, P.J.; Cagigal, M.P. Super-Gaussian apodization in ground based telescopes for high contrast coronagraph imaging. *Opt. Express* **2013**, *21*, 12744–12756. [CrossRef] [PubMed]
12. Rabbia, Y.; Gay, J.; Bascou, E. Achromatic phase shifters for nulling interferometry. In *International Conference on Space Optics—ICSO 2000*; Otrio, G., Ed.; International Society for Optics and Photonics, SPIE: Bellingham, WA, USA, 2017; Volume 10569, pp. 385–398. [CrossRef]

13. Goldsmith, H.D.K.; Cvetojevic, N.; Ireland, M.; Madden, S. Fabrication tolerant chalcogenide mid-infrared multimode interference coupler design with applications for Bracewell nulling interferometry. *Opt. Express* **2017**, *25*, 3038–3051. [CrossRef] [PubMed]
14. Collaboration, L.; Quanz, S.P.; Ottiger, M.; Fontanet, E.; Kammerer, J.; Menti, F.; Dannert, F.; Gheorghe, A.; Absil, O.; Airapetian, V.S.; et al. Large Interferometer For Exoplanets (LIFE): I. Improved Exoplanet Detection Yield Estimates for a Large Mid-Infrared Space-Interferometer Mission. *arXiv* **2015**, arXiv:2101.07500.
15. Scholl, M.S.; Paez, G. Cancellation of star light generated by a nearby star–planet system upon detection with a rotationally-shearing interferometer. *Infrared Phys. Technol.* **1999**, *40*, 357–365. [CrossRef]
16. Scholl, M.S. Infrared signal generated by a planet outside the solar system discriminated by a rotating rotationally-shearing interferometer. *Infrared Phys. Technol.* **1996**, *37*, 307–312. [CrossRef]
17. Bravo-Medina, B.; Strojnik, M.; Ipus, E. Comparison of Nulling Interferometry and Rotational Shearing Interferometry for Detection of Extrasolar Planets. In *Progress in Optomechatronic Technologies*; Springer: Berlin/Heidelberg, Germany, 2019; pp. 185–190.
18. Bravo-Medina, B.; Strojnik, M.; Kranjc, T. Feasibility of planet detection in two-planet solar system with rotationally-shearing interferometer. In *Infrared Remote Sensing and Instrumentation XXVII*; Strojnik, M., Arnold, G.E., Eds.; International Society for Optics and Photonics, SPIE: Bellingham, WA, USA, 2019; Volume 11128, pp. 75–86. [CrossRef]
19. Anglada-Escudé, G.; Amado, P.J.; Barnes, J.; Berdiñas, Z.M.; Butler, R.P.; Coleman, G.A.; de La Cueva, I.; Dreizler, S.; Endl, M.; Giesers, B.; et al. A terrestrial planet candidate in a temperate orbit around Proxima Centauri. *Nature* **2016**, *536*, 437–440. [CrossRef] [PubMed]
20. Thompson, A.R.; Moran, J.M.; Swenson, G.W., Van Cittert–Zernike Theorem, Spatial Coherence, and Scattering. In *Interferometry and Synthesis in Radio Astronomy*; Springer International Publishing: Cham, Switzerland, 2017; pp. 767–786. [CrossRef]
21. Strojnik, M.; Kirk, M. Telescopes. In *Fundamentals of Basic Optical Instruments*; CRC Press: Boca Raton, FL, USA, 2017; pp. 207–226.
22. Strojnik, M.; Bravo-Medina, B. Rotationally shearing interferometer for extra-solar planet detection: Preliminary results with a solar system simulator. *Opt. Express* **2020**, *28*, 29553–29561. [CrossRef] [PubMed]
23. Strojnik, M.; Bravo-Medina, B. Response of rotational shearing interferometer to a planetary system with two planets: Simulation. In *Modeling Aspects in Optical Metrology VII*; Bodermann, B., Frenner, K., Silver, R.M., Eds.; International Society for Optics and Photonics, SPIE: Bellingham, WA, USA, 2019; Volume 11057, pp. 8–18. [CrossRef]

Article

Automatic Generation of Aerial Orthoimages Using Sentinel-2 Satellite Imagery with a Context-Based Deep Learning Approach

Suhong Yoo, Jisang Lee , Junsu Bae, Hyoseon Jang and Hong-Gyoo Sohn *

School of Civil and Environmental Engineering, Yonsei University, Seodaemun-gu, Seoul 03722, Korea; swennoir@yonsei.ac.kr (S.Y.); ontheground@yonsei.ac.kr (J.L.); junsu510@yonsei.ac.kr (J.B.); hyoseon9026@yonsei.ac.kr (H.J.)
* Correspondence: sohn1@yonsei.ac.kr; Tel.: +82-2-2123-2809

Abstract: Aerial images are an outstanding option for observing terrain with their high-resolution (HR) capability. The high operational cost of aerial images makes it difficult to acquire periodic observation of the region of interest. Satellite imagery is an alternative for the problem, but low-resolution is an obstacle. In this study, we proposed a context-based approach to simulate the 10 m resolution of Sentinel-2 imagery to produce 2.5 and 5.0 m prediction images using the aerial orthoimage acquired over the same period. The proposed model was compared with an enhanced deep super-resolution network (EDSR), which has excellent performance among the existing super-resolution (SR) deep learning algorithms, using the peak signal-to-noise ratio (PSNR), structural similarity index measure (SSIM), and root-mean-squared error (RMSE). Our context-based ResU-Net outperformed the EDSR in all three metrics. The inclusion of the 60 m resolution of Sentinel-2 imagery performs better through fine-tuning. When 60 m images were included, RMSE decreased, and PSNR and SSIM increased. The result also validated that the denser the neural network, the higher the quality. Moreover, the accuracy is much higher when both denser feature dimensions and the 60 m images were used.

Keywords: aerial orthoimage; Sentinel-2; super-resolution; image simulation; residual U-Net

Citation: Yoo, S.; Lee, J.; Bae, J.; Jang, H.; Sohn, H.-G. Automatic Generation of Aerial Orthoimages Using Sentinel-2 Satellite Imagery with a Context-Based Deep Learning Approach. *Appl. Sci.* **2021**, *11*, 1089. https://doi.org/10.3390/app11031089

Academic Editor: Yang Dam Eo
Received: 1 January 2021
Accepted: 22 January 2021
Published: 25 January 2021

1. Introduction

Aerial imagery has been widely used for monitoring the surrounding environment due to its long history. Orthoimages created from aerial images can provide high-quality geospatial information taken at lower altitudes than satellite images. Continuously monitoring a rapidly changing environment requires reducing the observation period for a site. However, the tradeoff between spatial resolution and ground coverage prevents aerial images from covering a wide area. The role of aerial imagery has been gradually replaced by satellite imagery with its wide area coverage and regular repeat pass capabilities. Moreover, satellites equipped with multispectral sensors have enabled multiple applications such as resource management, urban research, facility mapping, and disaster monitoring.

The resolution of most of the current satellite images is still lower than that of aerial images. The price of commercially available high-resolution (HR) satellites has frequently hindered many researchers' progress in their projects. In most countries, including Korea, HR aerial orthoimages are provided to the public for free [1]. Furthermore, in the United States and the European Union, low- and medium-resolution satellite images are provided free of charge to users around the world. Research is needed to increase the resolution of mid- and low-resolution satellite images using freely available HR aerial images.

In the field of remote sensing, a visible improvement of image resolution primarily implies pan-sharpening. This method improves the resolution of low-resolution multispectral images using an HR panchromatic image. There are two typical approaches, one using

Intensity-Hue-Saturation (IHS) information [2] and one using principal component analysis (PCA) [3]. The primary concern for pan-sharpening is that it is applicable only when an HR panchromatic image is available. Consequently, the resolution of the pan-sharpened image cannot be higher than that of the input panchromatic image. With the recent development of deep learning techniques, studies to produce images with higher resolution than the input image have been conducted. Several studies using deep learning techniques have been published in the remote sensing community. Related studies can be largely divided into two usage categories: multiple sensors from one platform and multiple sensors from multiple platforms [4–10].

Improving the resolution of multispectral sensors from one (same) platform is usually performed by merging lower and higher multispectral images. Gargiulo et al. [5] enhanced a 20 m shortwave infrared (SWIR) image acquired by Sentinel-2 into a 10 m SWIR image. Similar to the pan-sharpening approach, the four-channel 10 m visible and NIR resolution images of Sentinel-2 were regarded as panchromatic. A shallow convolutional neural network (CNN) was constructed to improve the resolution of the SWIR image. The limitation of this study is that only the resolution of an SWIR image can be improved. Lanaras et al. [6] presented research results that can address this limitation. By constructing deep and dense neural network models, DSen2 and VDSen2, they improved the 20 m resolution of three red-edge and three SWIR images, two 60 m resolution images of water vapor, and 60 m SWIR mages of Sentinel-2 images into 10 m. They asserted that the model could be extended and improved from 20 m and 60 m to a 10 m resolution. However, the first category cannot produce images with higher resolution than the maximum resolution provided by the platform.

Another category is improving the resolution of multispectral sensors from multiple (different) platforms. Few studies have improved 30 m Landsat-8 satellite images to 10 m using Sentinel-2 images. Shao et al. [7] proposed the extended super-resolution convolutional neural network (ESRCNN) by blending Landsat-8 and Sentinel-2 data. They demonstrated the effectiveness of the deep learning-based fusion method for improving the resolution of Landsat-8 imagery. In their study, a performance comparison was performed using area-to-point regression kriging rather than other deep learning-based algorithms. Pouliot et al. [9] tested shallow and deep CNNs and confirmed that the deep CNN performed the same or better than the shallow CNN. The suggested algorithm demonstrated high-performance, but computational complexity and memory requirements could be problematic because the model is trained for each band.

After analyzing the previous studies, we found three common points. The first is that the use of deep neural networks is superior [6,8,9]. Tai et al. [8] analyzed the performance of each neural network by constructing shallow, deep, and very deep networks. They confirmed that the deeper the neural network, the higher the performance. Second, most neural networks have residual blocks and skip connections [6,8,10,11]. Consequently, the vanishing gradient problem can be alleviated, and the learning speed improved, even though the neural network is deeper. Third, the size of the input image inside the neural network is maintained until the last stage of the output, in contrast to neural networks for object detection and segmentation. Accordingly, the enlargement function to create the HR is only located in the final stage of neural networks using upsampling convolution layers or pixel shuffle algorithms [11]. Galar et al. [10] applied an enhanced deep super-resolution network (EDSR) to produce a 5 m resolution RapidEye RGB image with a 10 m resolution Sentinel-2 RGB image. They confirmed superior performance among super-resolution (SR) neural networks [11,12].

Studies so far have used neural networks of increasing resolution between satellite images. In this study, we propose a context-based ResU-Net to increase the resolution of Sentinel-2 imagery using 2.5 and 5.0 m downsampled aerial orthoimage acquired during the same period. For completing the tasks, the aerial orthoimages were first simulated by reconstructing a residual U-Net, which has advantages not only in constructing a deep and dense neural network but also in identifying adjacent contexts and the position of objects.

As a result of the experiments, we found that our neural network can express the aerial orthoimages' features and contexts well.

Training datasets were newly generated by using Sentinel-2 and aerial orthoimage. Sentinel-2 images, providing 10 m, 20 m, and 60 m resolution of multispectral bands, were utilized in this research. Sentinel-2 has the highest resolution and shortest revisit date among free satellite images. Since the advantage of obtaining many repeat pass images is a factor that can satisfy the objectives of this study, it was selected as input data. SR research is key to securing a high-resolution ground truth (GT), and aerial orthoimages are one of the most reliable and high-quality data. Therefore, aerial orthoimages with a similar acquisition date were utilized as the GT. Two types of aerial orthoimages were produced, 2.5 m and 5.0 m, as ground truth data downsampled from the original aerial orthoimagery. The data were used for testing two-times magnification (5.0 m based on 10 m) and challenged four-time magnification (2.5 m based on 10 m).

We tested the effect of using the lowest resolution 60 m image on the model and analyzed the model's influence when the feature dimensions are changed. In addition, the quality of our approach was investigated through the peak signal-to-noise ratio (PSNR), structural similarity index measure (SSIM), and root-mean-squared error (RMSE), a common approach in many SR studies.

Finally, we found that our model's performance in most metrics turned out to be better than that of EDSR. We also identified that incorporating 60 m resolution with 10 m resolution Sentinel-2 images outperforms the combination 10 m and 20 m resolution images. In addition, we confirmed that the denser feature dimensions have better performance. In particular, it could be a useful reference for related research as it predicts well even narrow roads that are difficult to identify with low-resolution satellite images.

2. Materials and Methods

2.1. Training Datasets Generation and Site Selection

2.1.1. Study Area

Daejeon City, located in the central part of the Korean peninsula, was selected as the study area. The city has an area of approximately 539 km^2 and is a transportation hub connecting the southern and northern regions. As depicted in Figure 1, most of the areas illustrate urban landscapes, where large and small buildings are clustered. Rice paddies/fields and mountainous areas are distributed in minimal areas. Middle areas, primarily covered with many complex environments such as urban buildings and roads, are the areas where the SR approach is challenging to apply.

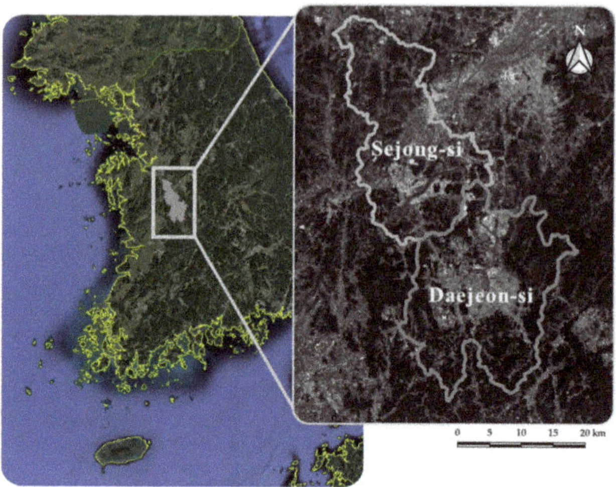

Figure 1. Location of Daejeon and Sejong depicted with Google map and Sentinel-2 images (band 2 image acquired on 2018/04/18).

Sejong City, Korea's administrative capital, is being developed into a city since 2012. The area of Sejong City is approximately 465 km^2, and most of the regions are still mountains and rice fields. However, due to construction, the impermeable layer is increasing rapidly every year. Daejeon City was selected to produce training datasets, and Sejong City was selected as a test site to analyze the generalization capabilities. Even if training samples and test samples are not overlapped, spatial autocorrelation within the same area cannot be avoided. Therefore, it was necessary to select an independent region with different characteristics.

2.1.2. Aerial Orthoimages

Aerial orthoimages were acquired in 2018, distributed free of charge under the leadership of the Korean government's aerial image acquisition and map production policy. Due to national security reasons, only 51 cm resolution images are provided to the public [1], and internally up to a 25 cm resolution is produced and used. We meticulously inspected the acquisition date of aerial images through the government orthoimage production manual and identified that aerial images were acquired over approximately one month, 21 April, 29 April, 5 May, and 26 May 2018, to cover the entire study area. The 51 cm orthoimages using the aerial triangulation method were provided through the government website. The final orthoimages downloaded are depicted in Figure 2.

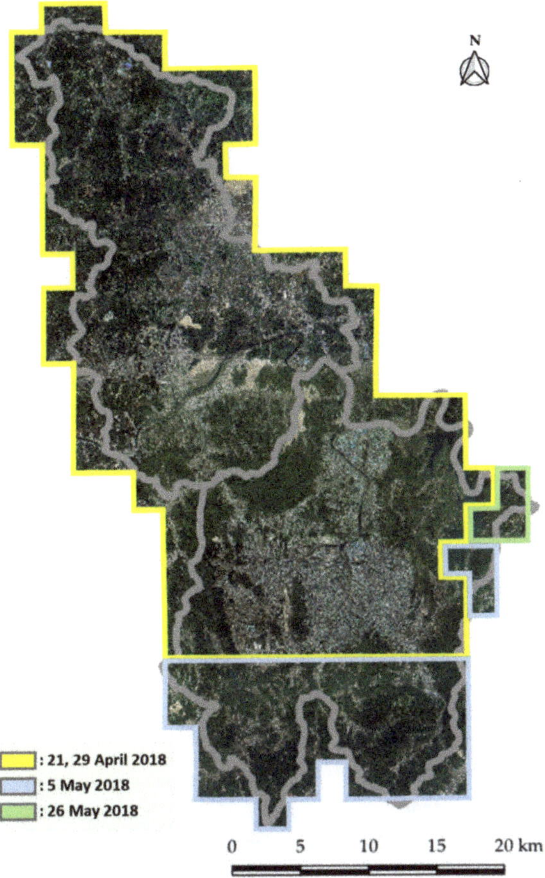

Figure 2. Aerial orthoimages over the study area with the acquisition date.

2.1.3. Sentinel-2A/B Satellite Imagery

Sentinel-2 is one of the satellites operated by the European Space Agency (ESA) and provides 13 multispectral bands with several different resolutions (10 m, 20 m, and 60 m). The imagery of Sentinel-2 has the highest resolution of 10 m among the current freely available for the general public. Accordingly, Sentinel-2 was selected for the study because it can provide much richer information than any other free satellite images. A short revisit period of five days is another strength of the Sentinel-2 imagery. Initially, the revisit period was ten days, but two satellites named Sentinel-2A and Sentinel-2B take images alternately, which reduces the revisit period to 5 days.

Sentinel-2 provides two types of images: (1) the L1C product, a top of atmosphere (TOA) reflectance image and (2) the L2A product, a bottom of atmosphere (BOA) reflectance image. The L2A product can overcome a significant difference in reflectivity, which varies for different acquisition times. Because aerial images are acquired at a much lower altitude than satellite images, it is better to use images with atmospheric correction. Because ESA provides the L1C product for images over the study area from 2018, all experimental images were converted to L2A through the Sen2cor tool of the Sentinel application platform (SNAP) software [9,13]. Twelve images (four 10 m, six 20 m, and two 60 m) with different spectral bands ranging from visible wavelength to SWIR were acquired. In some land classification studies, 60 m resolution images are not used because they are primarily for atmospheric correction [14,15]. However, we tested our approach with and without 60 m resolution imagery to consider whether additional atmospheric information is useful for training input images.

For matching the Sentinel-2 images acquired at the same time interval as the aerial orthoimages, data were searched through the Copernicus website, where the Sentinel series took all provided images [16]. We obtained both Sentinel-2A and 2B sensor images, which contain less cloud coverage, from the website. Searched images used in this research are listed in Table 1, and only band 2 images are depicted in Figure 3. All 10 m and 20 m images were used in training as defaults, with 60 m as optional. Because the datasets are acquired simultaneously with the aerial orthoimages, it was assumed that there were no significant topographic changes during the short period. Accordingly, listed datasets are used for all the following experiments.

Table 1. Sentinel-2 sensing start times used for the research.

Platform	Sensing Start Time
Sentinel-2A	2018/04/18 02:16:01
	2018/04/28 02:16:11
	2018/05/28 02:16:51
Sentinel-2B	2018/05/23 02:16:49
Total	4 images

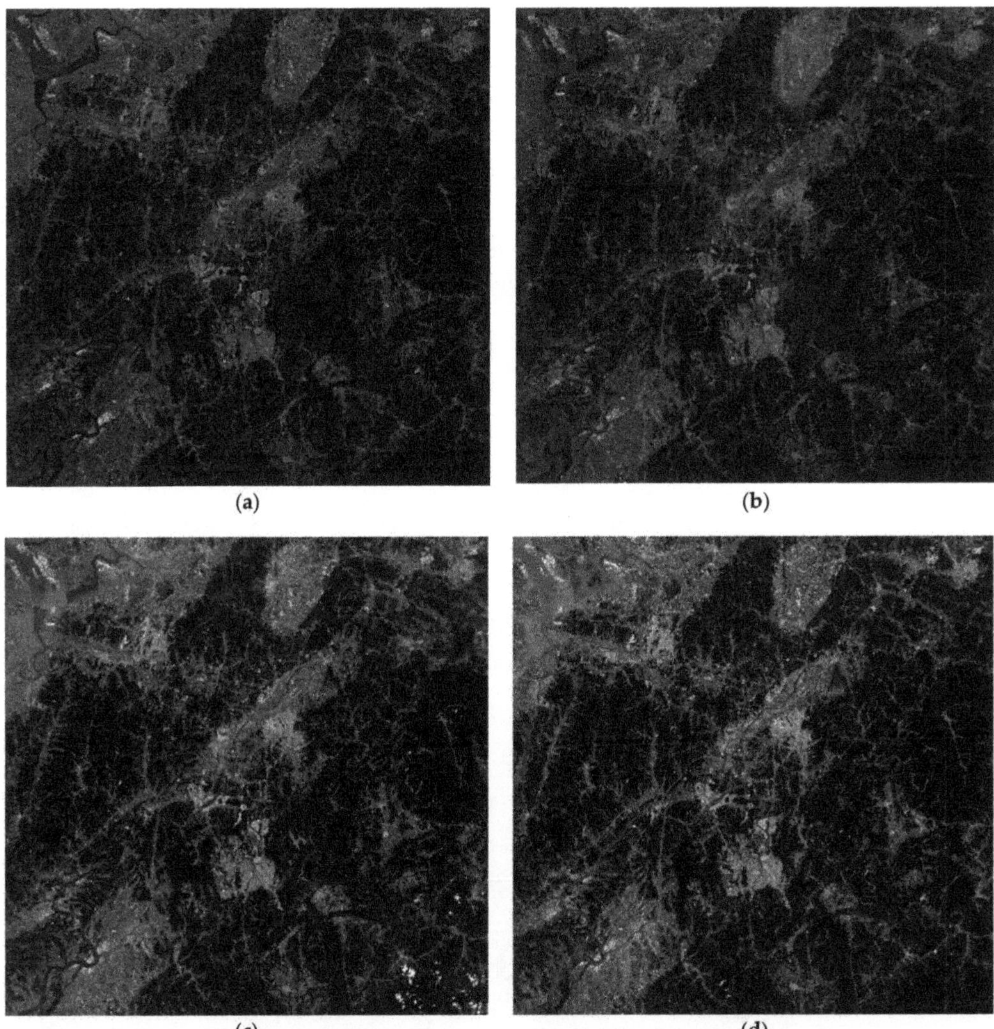

Figure 3. Acquired Sentinel-2 band number 2 (10 m) images: (**a**) Sentinel-2A 2018/04/18, (**b**) Setntinel-2A 2018/04/28, (**c**) Setntinel-2A 2018/05/28, (**d**) Setntinel-2B 2018/05/23.

2.1.4. Training Datasets Generation

The training datasets were preprocessed based on 2.5 m downsampled aerial orthoimages and 10 m Sentinel-2 satellite images. The first step was to transform both image sets into the same map projection system. All image sets in this study were projected into the Korea 2000 coordinate system (EPSG: 5186), corresponding to transverse mercator (TM) projection. The second step was to determine the size of training datasets based on 60 m Sentinel-2 images. After considering the computational efficiency of training processes, the 4×4 pixels image size was used, corresponding to 240×240 m^2 on the ground. For this configuration, the image size for 2.5 m and 5.0 m aerial orthoimages were 96×96 pixels and 48×48 pixels, respectively. For the same reason, the training image sizes of 10 m and 20 m resolution for Sentinel-2 were 24×24 pixels and 12×12 pixels, respectively.

Training samples and test samples were selected randomly within the study area but did not overlap for the Daejeon area. Through this process, 32,632 training samples (6527 for validation samples, 20% of the training samples) and 8156 test samples were produced. In addition, 39,204 test samples were generated for the Sejong area. Each set consisted of twelve Sentinel-2 images (4 for 10 m, 6 for 20 m, and 2 for 60 m) and two aerial photographs (1 for 2.5 m and 1 for 5.0 m), as depicted in Figure 4. A 5.0 m aerial orthoimage was used as the GT for 2× magnification of 10 m Sentinel-2 images and 2.5 m for 4× magnification of 10 m Sentinel-2 images.

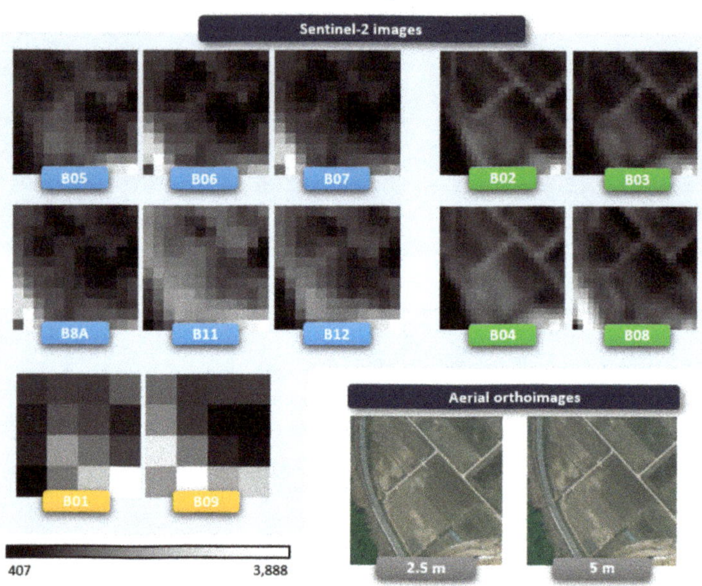

Figure 4. Example of training sets (Input: {B2 B3 B4 B8}: 10 m, {B5 B6 B7 B8A B11 B12}: 20 m, and {B1 B9}: 60 m, output: aerial orthoimages 2× for 5.0 m, 4× for 2.5 m).

2.2. Methodology

2.2.1. Context-Based ResU-Net

The latest research results indicate that the quality of SR increases as more convolution layers or deeper neural networks are assigned [6,8,9]. Most recent deep learning-based SR neural networks adopt this trend by maintaining the size of the input image until the output stage. The enlargement function to create HR is applied to the final stage [8,11,12]. The existing methodology was applied to our datasets with unsatisfactory results. It is speculated that different imaging geometry between aerial and space-borne sensors may lead to unsatisfactory results even with similar research methods. Because the aerial orthoimage contains more context information than the space-borne Sentinel-2 image, we determined that it would be critical to arrange context-preserving and deep and dense neural networks in the initial stage. The proposed architecture of the context-based ResU-Net for our study is depicted in Figure 5.

In our study, the residual U-Net proposed by Zhang et al. [17] was modified to maintain the context information and build deep neural networks. Batch normalization (BN) and ReLU activation functions are included in most of the steps. BN helps to solve gradient vanishing/exploding and overfitting caused by the deep neural network; it also improves accuracy [6,11]. The ReLU is used to remove the values below zero [6]. The encoder's role is to make the input image compact, and the decoder recovers the information to generate the final image. There is a path connecting the encoder and the decoder, and all convolution layers have a filter size of 3 × 3. The encoding path has three conv-depth blocks. Each block's stride was set to 2 instead of using downsampling layers to reduce the feature map's size in half. The decoding path has three conv-depth blocks to correspond to the encoder, and the size is increased through upsampling layers. End of the decoding path, a convolution layer is inserted to make feature dimensions as 3 with ReLU activation function for generating desired resolution similar to that of aerial orthoimage.

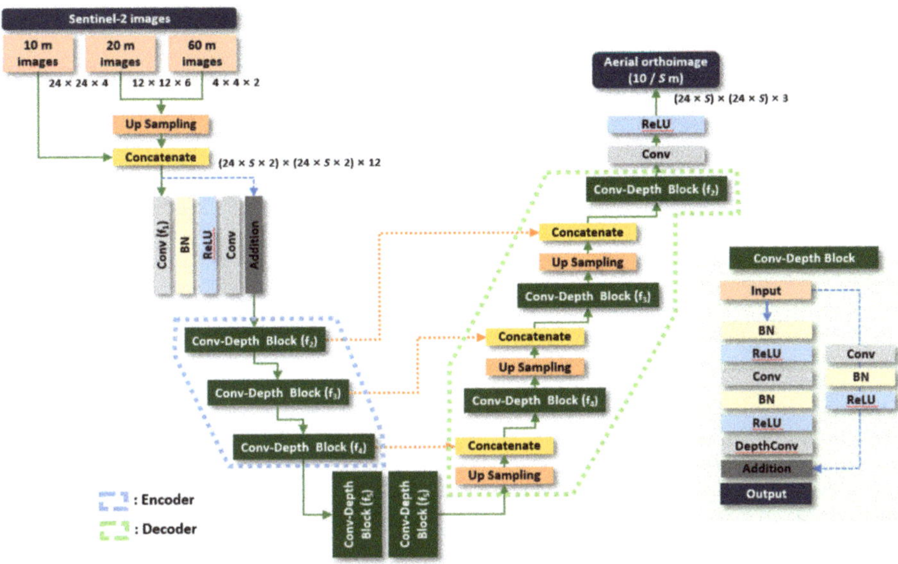

Figure 5. Context-based residual U-net for aerial images simulation (*S*: scale, *f*: feature dimension).

There are three major differences between the existing Residual U-Net and our network. First, the conv-depth block was included to reduce computation resources. It is known that depth-wise separable convolution (DepthConv) maintains performance while reducing the number of parameters [18]. As shown in Table 2, if a convolution layer is used instead of a DepthConv layer in our architecture, the number of parameters to be learned becomes larger. In addition, the difference in the number of parameters increased as the size of the feature dimensions increased. Moreover, we had encountered that the validation loss was jagged when only the convolution layer was used. On the contrary, the loss converges evenly with a lower value when using the DepthConv layer, as shown in Figure 6. When only the convolution layer was used, the loss at epoch 1 was 45,328.06, but the value was too large to be displayed on the graph, so only the corresponding value was clipped.

Table 2. Comparison of parameter numbers.

Feature Dimensions	Compositions	Number of Parameters	
		Trainable Parameters	Total Parameters
f_a	Using convolutional layer only	4,718,035	4,725,331
	Using DepthConv layer	3,159,955	3,167,251
f_b	Using convolutional layer only	18,845,091	18,859,683
	Using DepthConv layer	12,595,491	12,610,083
f_c	Using convolutional layer only	75,326,275	75,355,459
	Using DepthConv layer	50,293,315	50,322,499

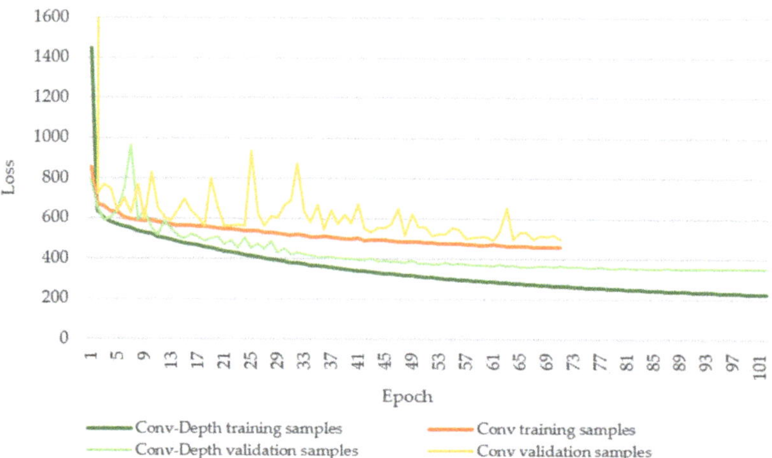

Figure 6. Loss convergence comparison: conv-depth block vs. conv only for context-based ResU-Net when using 60 m images with f_c feature dimensions (2×).

Second, upscaling was applied in the initial stage of the neural networks. The reason for changing the order like this is that the final prediction image becomes smoother or darker than that of the GT image when the image size is enlarged in the final stage as most of the other SR networks are arranged. In our networks, the scale (S) indicates an increasing factor of the original image. For example, the scale was set to $2S$ at the beginning and halved at the end, achieving a double improvement effect. Finally, the stride parameter was set to 2 to halve the image resolution.

Third, the feature dimension ($f = \{f_1, f_2, f_3, f_4, f_5\}$) was configured to increase gradually as the image size decreases, and the experiment was conducted in three groups: $f_a = \{16, 32, 64, 128, 256\}$, $f_b = \{32, 64, 128, 256, 512\}$, and $f_c = \{64, 128, 256, 512, 1024\}$. It was designed to analyze the predictive ability according to the size of feature dimensions.

2.2.2. Hyperparameter Optimization

The following hyperparameters were chosen to control the learning process. The related parameters were the optimizer, loss function, learning rate, batch size, and epoch. For an optimizer Adam optimizer for gradient descent was used in this study, reflecting many previous studies that this optimizer produced the best performance and had lower memory requirements than others [6,9,10,19]. The L1 loss function was used to minimize the error, which is the sum of all the absolute differences between the true value and the predicted value; it has been widely applied to SR neural networks [11,12,19]. For our study, we adopted the mean squared error (MSE) loss function instead of L1 because MSE had better results than L1. Finally, the mini-batch size was set to 32.

The learning rate gradually decreased as the epoch increased through a rate decay scheduler [18]. Consequently, the learning rate functions as an essential hyperparameter because it is dependent on the epoch parameter. Therefore, the epoch was adjusted between 30 and 180 to find the minimum loss; the experiment was repeated for each model. The initial learning rate was set to 5×10^{-4}. Early stopping criteria using validation datasets were also applied to avoid overfitting, and learning was stopped if the accuracy was not improved within 10 epochs. All programming was performed with Python-based Tensorflow nightly (2.5.0) GPU version, and learning was conducted using three graphics cards: two GeForce RTX-2080 Ti (11 GB VDRAM) and one RTX-3090 (24 GB VDRAM).

3. Results

Two experiments were conducted to evaluate our results: (1) whether to use 60 m images and (2) the effects of the feature dimension sizes. The EDSR neural network was also trained for comparison with our results. EDSR was selected as a comparison due to its excellent performance among the currently developed SR neural networks [12]. Lim et al. [11] designed both baseline and EDSR models. The difference between the two models is the number of residual blocks and the feature dimension. The baseline model is organized with 16 residual blocks and 64 feature dimensions, and EDSR is formed with 32 residual blocks and 256 feature dimensions. Both models are utilized for comparison, and all related training parameters were set as the author suggested. After training both neural networks with the same datasets, the results were evaluated with three metrics. PSNR and SSIM were used to evaluate the outcome—they are most frequently used as an evaluation index of SR deep learning research [11,12,20]. The RMSE used in some studies [6,9] was also included. The comparison of the final three metrics is summarized in Table 3 for Daejeon City and Table 4 for Sejong City, respectively. The scale parameters 2 and 4 refer to generating 5.0 m and 2.5 m aerial orthoimages, respectively.

In the case of Daejeon City, our context-based ResU-Net outperformed the baseline and EDSR models for all three metrics. For EDSR, even if the residual blocks and feature dimensions increased comparing with the baseline model, it is difficult to find performance improvement. In the case of Sejong City, in which independent testing was performed, our models performed better in two metrics except for RMSE.

Table 3. Quality evaluation results with set parameters for Daejeon City. Blue indicates the best results, and red indicates the second best.

Scale	Neural Networks	Use of 60 m	Feature Dimensions	Epoch	RMSE	PSNR	SSIM
2	Baseline (EDSR)	Y	64	120	22.9871	22.2210	0.4750
		N		100	23.1250	22.1883	0.4738
	EDSR	Y	256	70	22.5078	22.3772	0.4834
		N		100	21.9486	22.6070	0.4935
	Context-based ResU-Net (Ours)	Y	f_a	180	20.2371	23.3116	0.5010
			f_b		19.3775	23.6701	0.5233
			f_c		18.6816	23.9578	0.5437
		N	f_a		20.2274	23.3333	0.5005
			f_b		19.4900	23.6332	0.5234
			f_c		18.8066	23.8895	0.5439
4	Baseline (EDSR)	Y	64	180	24.6372	21.5330	0.3675
		N			24.7514	21.5305	0.3648
	EDSR	Y	256	30	26.4536	20.9607	0.3516
		N		40	26.7308	20.9715	0.3572
	Context-based ResU-Net (Ours)	Y	f_a	180	22.9141	22.1295	0.3770
			f_b		22.1827	22.3758	0.3888
			f_c		21.3574	22.6966	0.4101
		N	f_a	120	22.9897	22.1006	0.3774
			f_b		22.1444	22.3971	0.3912
			f_c		21.7778	22.5278	0.4018

Table 4. Quality evaluation results for Sejong City. Blue indicates the best results, and red indicates the second best.

Scale	Neural Networks	Use of 60 m	Feature Dimensions	RMSE	PSNR	SSIM
2	Baseline (EDSR)	Y	64	30.3523	19.3793	0.4034
		N		30.5805	19.3052	0.4009
	EDSR	Y	256	30.3873	19.4050	0.4086
		N		30.0645	19.4856	0.4110
	Context-based ResU-Net (Ours)	Y	f_a	30.4532	19.4819	0.4125
			f_b	30.8639	19.3829	0.4122
			f_c	30.4220	19.5121	0.4182
		N	f_a	30.5115	19.4902	0.4183
			f_b	30.3297	19.5689	0.4173
			f_c	30.4948	19.5190	0.4151
4	Baseline (EDSR)	Y	64	31.5568	18.9689	0.3287
		N		31.4554	18.9719	0.3270
	EDSR	Y	256	32.4436	18.7164	0.3188
		N		33.1779	18.5482	0.3250
	Context-based ResU-Net (Ours)	Y	f_a	31.3203	19.0943	0.3357
			f_b	31.5619	19.0178	0.3357
			f_c	31.6259	19.0225	0.3387
		N	f_a	31.8976	18.9556	0.3368
			f_b	31.4533	19.1072	0.3382
			f_c	31.3686	19.0990	0.3407

The image quality deteriorated as the magnification was enlarged, and the value of the evaluation metrics gradually deteriorated. Through fine-tuning, the inclusion of 60 m images performs better in two networks. When 60 m images were included, RMSE decreased, and PSNR and SSIM increased. This result demonstrates that the 60 m images have a positive impact on both networks.

The loss converged to a lower value if feature dimensions increased, as depicted in Figure 7. The result also validated that the denser the neural network, the higher the quality. Moreover, we found the accuracy is much higher when both denser feature dimensions and the 60 m images were used.

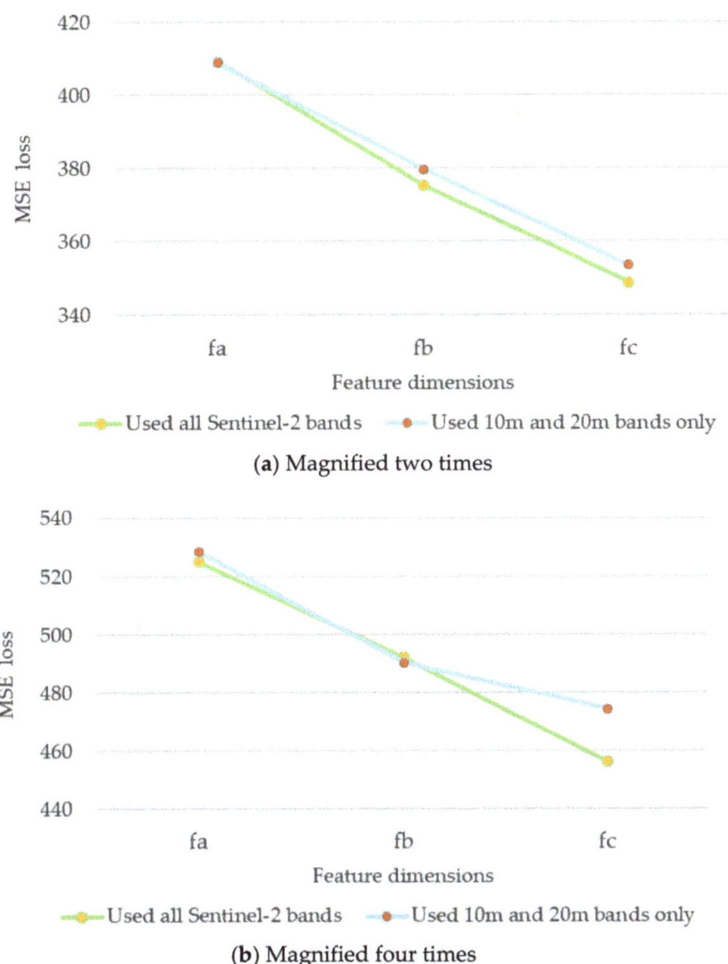

Figure 7. Convergence results according to each feature dimension.

For a visual comparison between EDSR and context-based ResU-Net, the prediction images are listed in Tables 5–9. Each table shows the predicted image of one representative input Sentinel-2 image per resolution (10 m, 20 m, and 60 m) for two scales, 2 and 4. The 2.5 m and 5.0 m aerial orthoimages are GT. The use of the 60 m Sentinel-2 images is shown in the second column. The predicted images of the baseline and EDSR model are shown in the third column. There is not much difference between the baseline and EDSR model, and only EDSR will be compared in the following. The predicted images of our context-based ResU-Net for three feature dimensions (f_a, f_b, f_c) are shown in the fourth column of each table. Generally, the prediction images between EDSR and context-based ResU-Net are visually similar when the feature dimension of context-based ResU-Net is f_a. For EDSR, even if the residual blocks and feature dimensions increase, no further improvement can be found. However, in our model, as networks become denser from f_a to f_c, it can be seen that the prediction images are getting close to GT.

Table 5. Predicted images of Sentinel-2 and corresponding GT image (paddy/road area).

Scale	Use of 60 m	Predicted Images			Input Images per Each Resolution (Sentinel-2)
		Baseline and EDSR		Context-Based ResU-Net (Ours)	
2	Yes	64	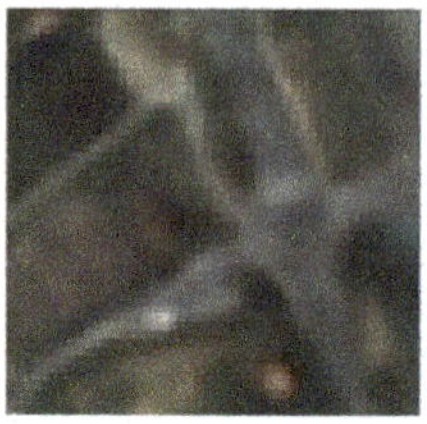	f_a	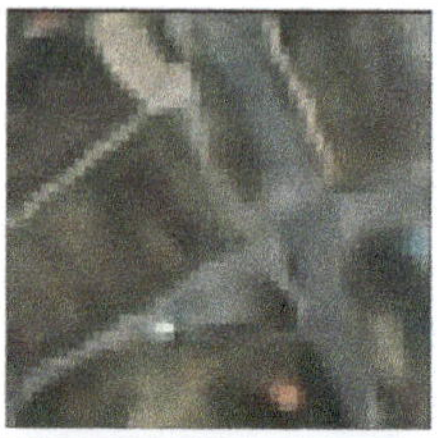  < Band 01 (60 m) >
				f_b	 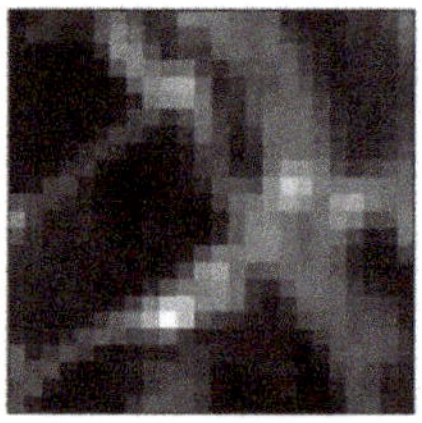 < Band 02 (10 m) >
		256	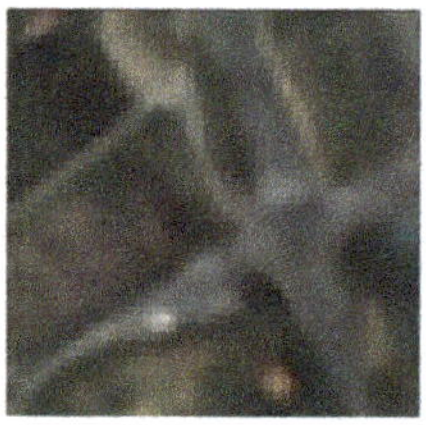	f_c	  < Band 05 (20 m) >
	No	64		f_a	
				f_b	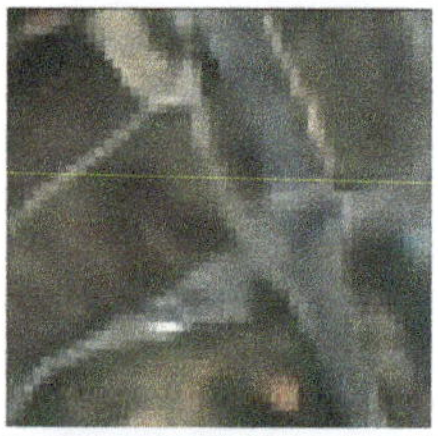 **GT image**
		256		f_c	  < Orthoimage (5.0 m) >

Table 5. *Cont.*

Scale	Use of 60 m	Predicted Images			Input Images per Each Resolution (Sentinel-2)
		Baseline and EDSR		Context-Based ResU-Net (Ours)	
4	Yes	64		f_a	< Band 01 (60 m) >
				f_b	< Band 02 (10 m) >
		256		f_c	< Band 05 (20 m) >
	No	64		f_a	GT image
				f_b	
		256		f_c	< Orthoimage (2.5 m) >

Table 6. Predicted images of Sentinel-2 and corresponding GT image (urban area).

Scale	Use of 60 m	Predicted Images		Input Images per Each Resolution (Sentinel-2)
		Baseline and EDSR	Context-Based ResU-Net (Ours)	
2	Yes	64	f_a	< Band 01 (60 m) >
			f_b	< Band 02 (10 m) >
		256	f_c	< Band 05 (20 m) >
	No	64	f_a	
			f_b	GT image
		256	f_c	< Orthoimage (5.0 m) >

Table 6. *Cont.*

Scale	Use of 60 m	Predicted Images			Input Images per Each Resolution (Sentinel-2)
		Baseline and EDSR		Context-Based ResU-Net (Ours)	
4	Yes	64		f_a	< Band 01 (60 m) >
				f_b	< Band 02 (10 m) >
		256		f_c	< Band 05 (20 m) >
	No	64		f_a	
				f_b	 GT image
		256		f_c	  < Orthoimage (2.5 m) >

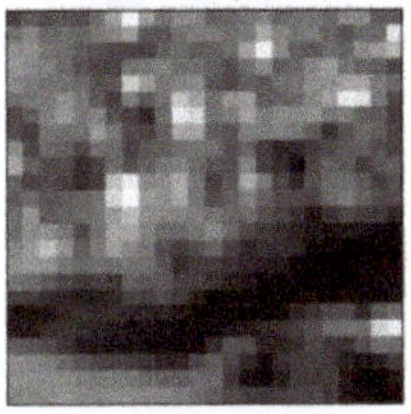

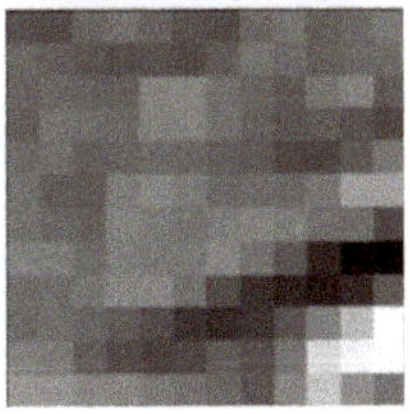

Table 7. Predicted images of Sentinel-2 and corresponding GT image (forest area).

Scale	Use of 60 m		Predicted Images: Baseline and EDSR	Predicted Images: Context-Based ResU-Net (Ours)		Input Images per Each Resolution (Sentinel-2)
2	Yes	64		f_a		< Band 01 (60 m) >
				f_b		< Band 02 (10 m) >
		256		f_c		< Band 05 (20 m) >
	No	64		f_a		GT image
				f_b		
		256		f_c		< Orthoimage (5.0 m) >

Table 7. *Cont.*

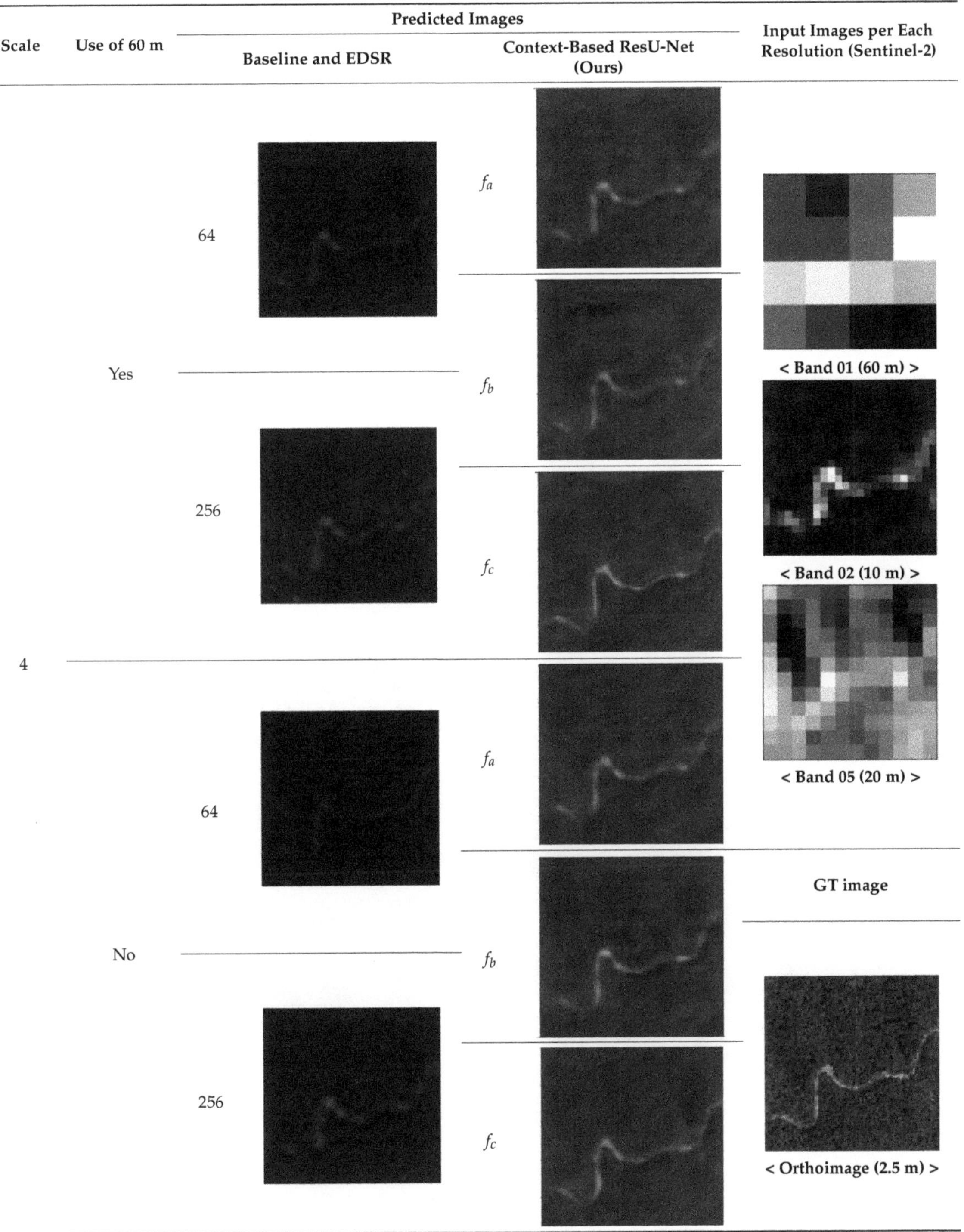

Scale	Use of 60 m	Predicted Images			Input Images per Each Resolution (Sentinel-2)
		Baseline and EDSR		Context-Based ResU-Net (Ours)	
4	Yes	64		f_a	< Band 01 (60 m) >
				f_b	< Band 02 (10 m) >
		256		f_c	< Band 05 (20 m) >
	No	64		f_a	
				f_b	GT image
		256		f_c	< Orthoimage (2.5 m) >

Table 8. Predicted images of Sentinel-2 and corresponding GT image (urban/forest area).

Scale	Use of 60 m	Predicted Images			Input Images per Each Resolution (Sentinel-2)
		Baseline and EDSR		Context-Based ResU-Net (Ours)	
2	Yes	64		f_a	< Band 01 (60 m) >
				f_b	< Band 02 (10 m) >
		256		f_c	< Band 05 (20 m) >
	No	64		f_a	
				f_b	**GT image**
		256		f_c	< Orthoimage (5.0 m) >

Table 8. *Cont.*

Scale	Use of 60 m	Predicted Images			Input Images per Each Resolution (Sentinel-2)
		Baseline and EDSR		Context-Based ResU-Net (Ours)	
4	Yes	64		f_a	  < Band 01 (60 m) >
				f_b	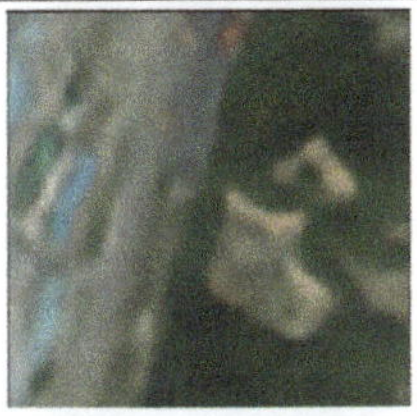 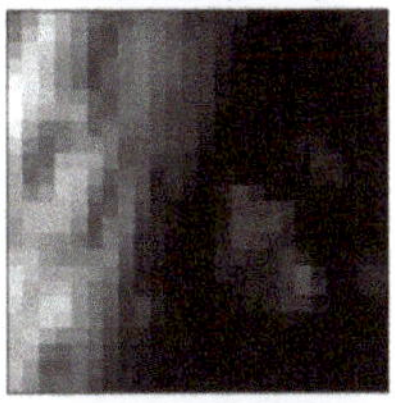 < Band 02 (10 m) >
		256		f_c	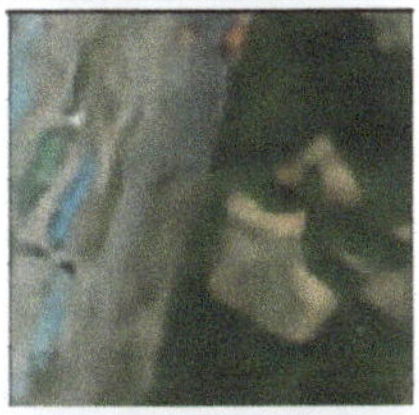  < Band 05 (20 m) >
	No	64		f_a	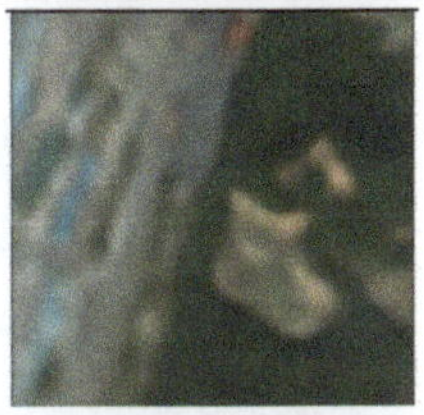 **GT image**
				f_b	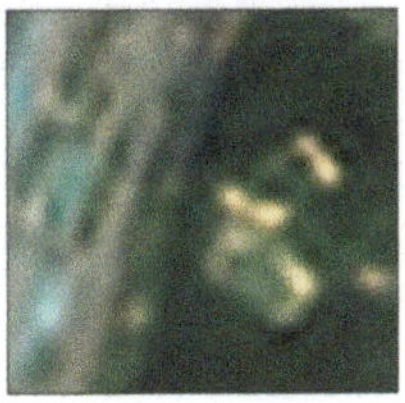   < Orthoimage (2.5 m) >
		256		f_c	

Table 9. Predicted images of Sentinel-2 and corresponding GT image (urban/road area).

Scale	Use of 60 m		Predicted Images			Input Imagesper Each Resolution (Sentinel-2)
			Baseline and EDSR		Context-Based ResU-Net (Ours)	
2	Yes	64		f_a		< Band 01 (60 m) >
				f_b		< Band 02 (10 m) >
		256		f_c		< Band 05 (20 m) >
	No	64		f_a		GT image
				f_b		
		256		f_c		< Orthoimage (5.0 m) >

Table 9. *Cont.*

Scale	Use of 60 m	Predicted Images: Baseline and EDSR	Predicted Images: Context-Based ResU-Net (Ours)	Input Imagesper Each Resolution (Sentinel-2)
4	Yes	64	f_a	< Band 01 (60 m) >
			f_b	< Band 02 (10 m) >
		256	f_c	< Band 05 (20 m) >
	No	64	f_a	
			f_b	GT image
		256	f_c	< Orthoimage (2.5 m) >

Observing the boundaries of an object reveals the difference between the two methods. For EDSR, when the image was enlarged four times, the overall boundary of each object remained similar or smoother than that of two-time enlargement-causing the prediction images to look blurry. For context-based ResU-Net, boundaries of each object became more distinct as the feature dimensions increased in density regardless of the enlargement scale. When the feature dimension reached its maximum size, the boundaries of the object became sharpest. Consequently, the visibility of all images improved.

Some differences were found between the two models. EDSR predicts a darker image, especially in forest areas (Table 7), and it produced a blurry image compared to ours, as shown in Tables 5–9. Interestingly, the result of context-based ResU-Net predicts even urban shadows well in the densest feature dimension f_c, which are not even expressed in EDSR as shown in Table 6. It was also identified that our model generally trained the boundaries of objects better. In particular, road boundaries are well preserved even the width is narrower than the 10 m resolution Sentinel-2 image. The result implies that the recognition of the object of concern, such as the road, can be possible by using predicted Sentinel-2 imagery, as shown in Tables 5 and 7. The road boundaries become clear as feature dimension becomes denser, but it seems that some attention needs to be paid to the shape of the road for the inclusion of 60 m Sentinel-2 imagery. Some of the road boundaries were visually curved when the 60 m Sentinel-2 images were included, but they were straight when the 60 m image was not included. It can be said that there exists a tradeoff between the value of metrics and visualization when the road boundaries are concerned.

4. Discussion

A study was conducted to produce 2.5 m and 5.0 m resolution imagery with 10 m Sentinel-2 satellite images using aerial orthoimage as a ground truth. For this, training samples were produced by acquiring Sentinel-2 satellite images and aerial orthoimages over the same area and period. The training samples were used to simulate 2.5 m and 5.0 m aerial orthoimages. For quality check and general applicability of our neural network, additional test samples in an independent region were utilized. For producing better-simulated images, a new context-based neural network was proposed and compared with the existing neural network. Our context-based ResU-Net generally outperformed the baseline and EDSR for all three metrics, both in training samples and test samples. We believe that this is because conv-depth blocks helped the stability of our model. In any case, the utility of our model for successfully predicting narrow roads will be very high. Meanwhile, in order to improve the performance compared to the present, the obstacles to be solved were speculated as follows:

First, the effect of shadows in HR aerial images was significant. The Sentinel-2 images were acquired with a low-resolution at high altitude, whereas aerial images were acquired with HR at low altitude. Even in the same area, when images were acquired at a low altitude, the effect of shadows was much more prominent than at a high altitude. Because most of our study area included urban landscapes, the effect of shadows on HR images was much greater than for high altitude images. The original 51 cm aerial orthoimage was resampled to obtain GT using bilinear interpolation. During the bilinear interpolation process to create GT from the original 51 cm aerial orthoimage, the effect shadow smeared into other features and worsened the SSIM metric.

Second, there existed the effects of color correction during the composition of aerial orthoimages. The primary purpose of aerial orthoimages distributed by the Korean government is to produce a visually attractive map for the general public. We speculated that the original reflectance information had been corrected to make the orthoimage more pleasing, leading to potentially difficult and inaccurate training due to the use of aerial orthoimages.

For this study, it is essential that both aerial orthoimages and satellite images must be taken at a similar period of time. Recently, some countries have provided aerial orthoimages, so if researchers can check the acquisition date of aerial orthoimages, we expect that our research results can be utilized.

In future research, steps for shadow identification and shadow removal must be included based on deep learning, especially when using the HR aerial images as training sets. In the remote sensing community, CNN-based SR research is ongoing. However, several studies have tried to combine images obtained from multiple sensors to produce new images. We believe that the method and results presented in this study can contribute new insights for researchers performing similar studies.

Author Contributions: S.Y. and H.-G.S. were the leading directors of this research. S.Y. designed the overall research plan and programmed them for experiments, while J.L. programmed training data generation and augmentation. J.B. and H.J. performed the preprocessing of aerial and satellite images. All authors have read and agreed to the published version of the manuscript.

Funding: This research was supported by a grant (no. 20009742) from the Disaster-Safety Industry Promotion Program funded by the Ministry of Interior and Safety (MOIS, Korea).

Institutional Review Board Statement: Not applicable.

Informed Consent Statement: Not applicable.

Conflicts of Interest: The authors declare no conflict of interest.

References

1. National Geographic Information Institute National Territory Information Platform. Available online: http://map.ngii.go.kr/mn/mainPage.do (accessed on 31 December 2020).
2. Rahmani, S.; Strait, M.; Merkurjev, D.; Moeller, M.; Wittman, T. An adaptive IHS pan-sharpening method. *IEEE Geosci. Remote Sens. Lett.* **2010**, *7*, 746–750. [CrossRef]
3. Ghadjati, M.; Moussaoui, A.; Boukharouba, A. A novel iterative PCA–based pansharpening method. *Remote Sens. Lett.* **2019**, *10*, 264–273. [CrossRef]
4. Liebel, L.; Körner, M. Single-image super resolution for multi-spectral remote sensing data using convolutional neural networks. *ISPRS-Int. Arch. Photogramm. Remote Sens. Spat. Inf. Sci.* **2016**, *41*, 883–890. [CrossRef]
5. Gargiulo, M.; Mazza, A.; Gaetano, R.; Ruello, G.; Scarpa, G. A CNN-Based Fusion Method for Super-Resolution of Sentinel-2 Data. In Proceedings of the IGARSS 2018-2018 IEEE International Geoscience and Remote Sensing Symposium, Valencia, Spain, 22–27 July 2018; pp. 4713–4716.
6. Lanaras, C.; Bioucas-Dias, J.; Galliani, S.; Baltsavias, E.; Schindler, K. Super-resolution of Sentinel-2 images: Learning a globally applicable deep neural network. *ISPRS J. Photogramm. Remote Sens.* **2018**, *146*, 305–319. [CrossRef]
7. Shao, Z.; Cai, J.; Fu, P.; Hu, L.; Liu, T. Deep learning-based fusion of Landsat-8 and Sentinel-2 images for a harmonized surface reflectance product. *Remote Sens. Environ.* **2019**, *235*, 111425. [CrossRef]
8. Tai, Y.; Yang, J.; Liu, X. Image Super-Resolution via Deep Recursive Residual Network. In Proceedings of the IEEE Conference on Computer Vision and Pattern Recognition Workshops, Honolulu, USA, 21–26 July 2017; pp. 3147–3155.
9. Pouliot, D.; Latifovic, R.; Pasher, J.; Duffe, J. Landsat super-resolution enhancement using convolution neural networks and Sentinel-2 for training. *Remote Sens.* **2018**, *10*, 394. [CrossRef]
10. Galar, M.; Sesma, R.; Ayala, C.; Aranda, C. Super-Resolution for Sentinel-2 Images. In Proceedings of the International Archives of the Photogrammetry, Remote Sensing and Spatial Information Sciences—ISPRS Archives, Nanjing, China, 25–27 October 2019; Volume 42, pp. 95–102.
11. Lim, B.; Son, S.; Kim, H.; Nah, S.; Mu Lee, K. Enhanced Deep Residual Networks for Single Image Super-Resolution. In Proceedings of the IEEE Conference on Computer Vision and Pattern Recognition Workshops, Honolulu, USA, 21–26 July 2017; pp. 136–144.
12. Zhang, Y.; Li, K.; Li, K.; Wang, L.; Zhong, B.; Fu, Y. Image Super-Resolution Using Very Deep Residual Channel Attention Networks. In Proceedings of the European Conference on Computer Vision (ECCV), Munich, Germany, 8–14 September 2018; pp. 286–301.
13. Thanh Noi, P.; Kappas, M. Comparison of random forest, k-nearest neighbor, and support vector machine classifiers for land cover classification using Sentinel-2 imagery. *Sensors* **2018**, *18*, 18. [CrossRef] [PubMed]
14. Wang, Q.; Shi, W.; Li, Z.; Atkinson, P.M. Fusion of Sentinel-2 images. *Remote Sens. Environ.* **2016**, *187*, 241–252. [CrossRef]
15. Gašparović, M.; Jogun, T. The effect of fusing Sentinel-2 bands on land-cover classification. *Int. J. Remote Sens.* **2018**, *39*, 822–841. [CrossRef]
16. European Space Agency (ESA) Copernicus. Available online: https://scihub.copernicus.eu/dhus/#/home (accessed on 31 December 2020).
17. Zhang, Z.; Liu, Q.; Wang, Y. Road extraction by deep residual u-net. *IEEE Geosci. Remote Sens. Lett.* **2018**, *15*, 749–753. [CrossRef]

18. Chen, L.-C.; Zhu, Y.; Papandreou, G.; Schroff, F.; Adam, H. Encoder-Decoder with Atrous Separable Convolution for Semantic Image Segmentation. In Proceedings of the European Conference on Computer Vision (ECCV), Munich, Germany, 8–14 September 2018; pp. 801–818.
19. Sun, Y.; Xu, W.; Zhang, J.; Xiong, J.; Gui, G. *Super-Resolution Imaging Using Convolutional Neural Networks. Lecture Notes in Electrical Engineering*; Springer: Berlin/Heidelberg, Germany, 2020; Volume 516, pp. 59–66.
20. Dong, C.; Loy, C.C.; He, K.; Tang, X. Learning a Deep Convolutional Network for Image Super-Resolution. In *Proceedings of the European Conference on Computer Vision, Zurich, Switzerland, 6–12 September 2014*; Springer: Berlin/Heidelberg, Germany, 2014; pp. 184–199.

Article

Kinematic In Situ Self-Calibration of a Backpack-Based Multi-Beam LiDAR System

Han Sae Kim [1], Yongil Kim [2], Changjae Kim [1] and Kang Hyeok Choi [3,*]

[1] Department of Civil and Environmental Engineering, Myongji University, 116 Myongji-ro, Cheoin-gu, Yongin 17058, Gyeonggi-do, Korea; hanswfg@snu.ac.kr (H.S.K.); cjkim@mju.ac.kr (C.K.)
[2] Department of Civil & Environmental Engineering, Seoul National University, 599 Gwanak-ro 1, Gwanak-gu, Seoul 08826, Korea; yik@snu.ac.kr
[3] Lyles School of Civil Engineering, Purdue University, 610 Purdue Mall, West Lafayette, IN 47907, USA
* Correspondence: choi663@purdue.edu; Tel.: +1-213-374-6043

Abstract: Light Detection and Ranging (LiDAR) remote sensing technology provides a more efficient means to acquire accurate 3D information from large-scale environments. Among the variety of LiDAR sensors, Multi-Beam LiDAR (MBL) sensors are one of the most extensively applied scanner types for mobile applications. Despite the efficiency of these sensors, their observation accuracy is relatively low for effective use in mobile mapping applications, which require measurements at a higher level of accuracy. In addition, measurement instability of MBL demonstrates that frequent re-calibration is necessary to maintain a high level of accuracy. Therefore, frequent in situ calibration prior to data acquisition is an essential step in order to meet the accuracy-level requirements and to implement these scanners for precise mobile applications. In this study, kinematic in situ self-calibration of a backpack-based MBL system was investigated to develop an accurate backpack-based mobile mapping system. First, simulated datasets were generated for the experiments and tested in a controlled environment to inspect the minimum network configuration for self-calibration. For this purpose, our own-developed simulator program was first utilized to generate simulation datasets with various observation settings, network configurations, test sites, and targets. Afterwards, self-calibration was carried out using the simulation datasets. Second, real datasets were captured in a kinematic situation so as to compare the calibration results with the simulation experiments. The results demonstrate that the kinematic self-calibration of the backpack-based MBL system could improve the point cloud accuracy with Root Mean Square Error (RMSE) of planar misclosure up to 81%. Conclusively, in situ self-calibration of the backpack-based MBL system can be performed using on-site datasets, reaching the higher accuracy of point cloud. In addition, this method, by performing automatic calibration using the scan data, has the potential to be adapted to on-line re-calibration.

Keywords: multi-beam LiDAR; in situ self-calibration; mobile mapping system; 3D point cloud; backpack-based mapping

Citation: Kim, H.S.; Kim, Y.; Kim, C.; Choi, K.H. Kinematic In Situ Self-Calibration of a Backpack-Based Multi-Beam LiDAR System. *Appl. Sci.* **2021**, *11*, 945. https://doi.org/10.3390/app11030945

Academic Editor: Stephen Grebby
Received: 27 November 2020
Accepted: 19 January 2021
Published: 21 January 2021

1. Introduction

Over the past few years, significant developments of laser scanning technology have increased the feasibility of acquiring large amounts of accurate geometric 3D data. The demand for 3D observation has increased as well with the development of automatic digital image analysis, namely artificial intelligence. In this context, laser scanners have become a fundamental means to acquire 3D information in a manner that is effective enough to satisfy the growing demand in this field [1]. This trend is particularly relevant in civil engineering, robotics, and computer vision, due to those fields' high usage of laser scanners as routine measurement techniques for applications such as 3D modeling and mapping [2,3], efficient building management [4,5], and the transformation of structural health monitoring [6,7]. Extending the application of laser scanners from geomatics to these domains has resulted in the emergence of new areas for potential implementation, which requires the development of more efficient and accurate acquisition systems [8]. More recently, Light Detection and Ranging (LiDAR) sensors have become more portable, compact, and readily available to be extended to mobile applications. The advancement of modern technology has expanded the use of LiDAR sensors to a variety of applications including mobile mapping, surveying, and autonomous vehicles. In particular, Multi-Beam LiDAR (MBL) is one of the most extensively used and favored sensors for mobile applications, because the sensors are relatively lightweight, compact, and cheap. MBL, manufactured by Velodyne LiDAR, has been widely used both in research and the industrial field [9–17]. Particularly, backpack-based 3D mapping using Velodyne LiDAR has been a popular mobile application. Leica Pegasus [18], Viametris bMS3D [19], GreenValley LiBackpack [20], and Gexcel Heron [21] are examples of the commercial backpack-based mobile mapping system. These solutions use two Velodyne VLP-16 units to perform odometry for georeferencing and 3D mobile mapping including Simultaneous Localization and Mapping (SLAM). Along with the growing need for 3D information, these commercial backpack-based mapping systems were released to match the demand for quick data acquisition and fieldwork planning; however, there are drawbacks that require further improvement in terms of accuracy and cost [22]. In this context, developing an accurate and affordable backpack-based mapping system is still an interest in this field.

For successful and effective mobile LiDAR scanning, two fundamental issues need to be addressed: georeferencing and sensor calibration. Georeferencing is the conversion of a local coordinate system to one global coordinate system combining all point clouds into the same coordinate system. Sensor calibration removes systematic errors inherent to sensors and can accomplish quality assurance to maximize the accuracy of observation. These two key processes are not independent, which means that the more accurate the sensor system that is built, the more accurate georeferencing becomes, so that the overall accuracy of surveying can be improved. Therefore, establishing an accurate mobile LiDAR sensor system is the most significant step. Even though MBL provides a cost-efficient and portable option, their observation accuracy is lower than conventional Terrestrial Laser Scanners (TLS) in general. They include systematic errors, which can affect the overall accuracy of the scanned data. Since each mechanically designed laser measures the range by time-of-flight system and encoder angle, the point cloud inevitably contains systematic errors in range and angle measurements with respect to each laser. These systematic errors can cause translations and rotations in the point cloud data. As a result, for precise mobile mapping and surveying, the overall accuracy of the point cloud data needs to be improved [23].

LiDAR self-calibration can remove the systematic errors inherent in the sensor and thus improve the overall accuracy of point cloud by reducing the Root Mean Square Error (RMSE) associated with registration and check points [24]. It also can reduce the need for point cloud outlier removal as a post-processing step. In the case of MBL, various studies in the literature have performed self-calibration using modified manufacturer-based calibration parameters. They also have confirmed the potential of applying these sensors as a basis for obtaining a highly accurate mobile mapping platform. A calibration of Velodyne HDL-64E, which is Velodyne's first generation of MBL consisting of 64 laser channels, can be found in the literature. Static calibration of Velodyne HDL-64E using plane-based targets achieved a 3D RMSE up to 0.013 m [25], while optimization-based calibration showed standard deviations of planar data from 0.006 to 0.037 m [26]. Moreover, minimizing the discrepancies between the point cloud and pattern planes attained 0.0156 m of point cloud accuracy [27]. In addition to the installed target-based approaches, the static on-site re-calibration approach using planes accomplished 0.013 m of planar misclosure [28]. Besides, the kinematic calibration of HDL-64E on a moving vehicle attained 0.023 m of planar RMS residuals [29]. Velodyne HDL-32E mounted on a vehicle, which consists of 32 laser channels, was also calibrated using cylinder-based self-calibration and improved the accuracy level to 0.008 m in static mode and 0.014 m in kinematic mode [30]. The calibration of the most recent generation of MBL by Velodyne, VLP-16, showed 0.025 m of planar RMSE residuals [31]. However, the system parameters may still be inconsistent, even after self-calibration, due to the instability inherent in the scanning system. Temporal stability analysis of an MBL demonstrated that the measurement stability is slightly higher than the quantization level, which stresses the need for periodic re-calibration of the LiDAR sensor to maintain a high level of accuracy [32]. Since measurement stability analysis showed inconsistency in range observation, there is a chance that the calibration parameters might change during data acquisition. As a result, periodic in situ calibration should be performed to increase and maintain the heightened overall accuracy of the point cloud for a backpack-based MBL system. This leads to our objective of the study, which is the kinematic self-calibration method that can be performed continually during the data acquisition.

In this respect, this study aimed to perform kinematic in situ self-calibration of a backpack-based MBL system for the purposes of easy, efficient and frequent periodical self-calibration prior to data acquisition. First, self-calibration was conducted with simulation datasets to examine the minimum network configuration for the in situ self-calibration of backpack-based MBL system. Second, based on the analysis from the simulation experiments, reals datasets were acquired using our own backpack system. The accuracy of the results was analyzed by investigating planar misclosure after the adjustment, the correlations between parameters, measurement residuals, and the standard deviation of the estimated parameters. The remainder of this study is organized as follows. Section 2 presents the configuration and specifications of the sensor system used in this study. Section 3 covers the mathematical models, which are an observation model, a systematic error model, a functional model, and a least squares solution for the adjustment. Section 4 outlines the experimental set-ups and the calibration datasets for the investigation of the minimum network configuration and observation requirements. It also includes the results of the experiments and accuracy analysis. Section 5 provides a discussion of the proposed method in terms of accuracy and benefits. In conclusion, Section 6 summarizes the findings of the study along with possible future work.

2. Backpack-Based Sensor System

The backpack-based MBL system used in this study consists of two LiDAR sensors, six optical cameras, and a Global Navigation Satellite System (GNSS)/Inertial Navigation System (INS) (Figure 1). The configuration of the involved sensors was determined to have minimum occlusion and maximum Field of View (FOV). The sensors are all integrated into a core computing system, which receives all sensor data and synchronizes into Universal Time Coordinated (UTC) timestamps. The Inertial Measurement Unit (IMU) was not described in Figure 1 since the unit is embedded in the computing system.

2.1. LiDAR Sensors

Two Velodyne VLP-16 (Figure 2) were mounted on the backpack system. Since its release in 2014, VLP-16 has been extensively utilized in mobile applications for both research and industry. The specifications of VLP-16 are summarized in Table 1. The sensor consists of 16 pairs of simultaneously rotating laser emitters and receivers within a compact sensor pod, and each laser has a fixed vertical angle of 2° resolution. The rotation rate varies from 5 to 20 Hz, with a set default value of 10 Hz, which gives 0.2° of horizontal angular resolution. Based on the default settings, VLP-16 rotates every 0.1 s and acquires approximately 28,800 points in each scan.

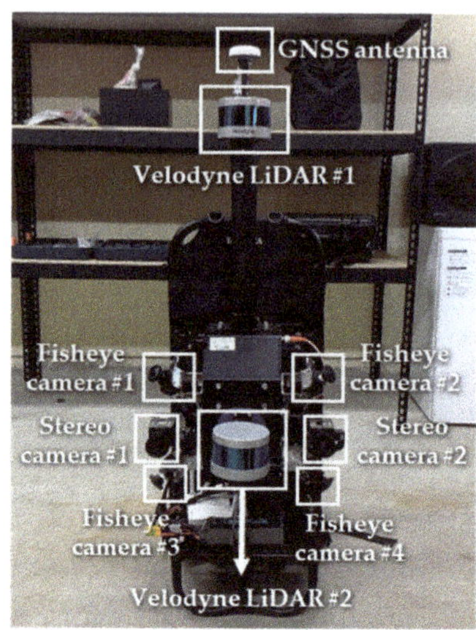

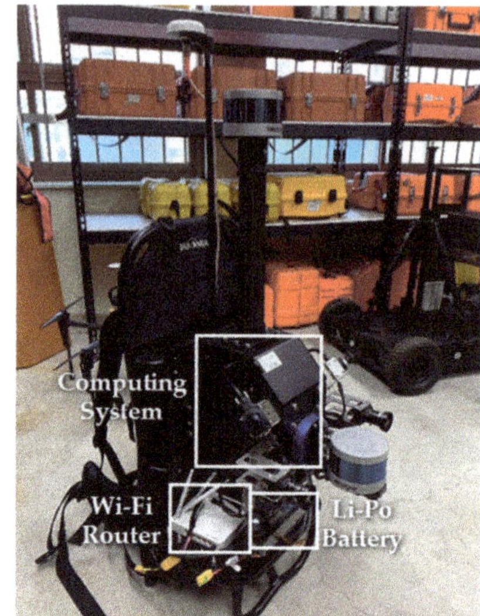

Figure 1. Backpack-based Multi-Beam Light Detection and Ranging (LiDAR) (MBL) sensor system.

Figure 2. Velodyne VLP-16.

Table 1. Specifications of Velodyne VLP-16.

Specification	
Channels	16 lasers
Range	Up to 100 m
Range Accuracy	Up to ±3 cm
FOV (Vertical)	+15.0° to −15.0° (30.0°)
Angular Resolution (Vertical)	2.0°
FOV (Horizontal)	360°
Angular Resolution (Horizontal)	0.1°–0.4°
Rotation Rate	5 Hz–20 Hz

2.2. Inertial Sensor

The Trimble APX-15 Unmanned Aerial Vehicle (UAV) was mounted for GNSS/INS. The Trimble APX-15 UAV (Figure 3) is an efficient GNSS-inertial solution for small UAVs. It weighs 60 g and is light enough to attach onto the backpack system. With GNSS signal integrated, APX-15 gives position, roll, pitch and heading output in 100 Hz, and IMU data in 200 Hz. This enables an accurate direct georeferencing of various sensor data. Detailed specifications of Trimble APX-15 UAV are described in Table 2.

Figure 3. Trimble APX-15 Unmanned Aerial Vehicle (UAV) single board Global Navigation Satellite System (GNSS)-inertial solution.

Table 2. Specifications of Trimble APX-15 UAV.

	Specification			
Size (mm)	67 L × 60 W × 15 H			
Weight	60 g			
IMU data rate	200 Hz			
	SPS	**DGPS**	**RTK**	**Post-Processed**
Position (m)	1.5–3.0	0.5–2.0	0.02–0.05	0.02–0.05
Velocity (m/s)	0.05	0.05	0.02	0.015
Roll & Pitch (deg)	0.04	0.03	0.03	0.025
True Heading (deg)	0.30	0.28	0.18	0.080

2.3. Digital Cameras

Four fisheye lens cameras and two perspective cameras were mounted on the backpack system. The model of the fisheye lens is Sunnex DSL315 and the camera body is Chameleon CM3-U3-32S4C (Figure 4). The camera has no shutter button, operating by receiving signals from a computing system. Detailed specifications of the fisheye camera are described in Table 3. Stereo cameras are also built in the backpack system. The perspective lens model is KOWA LM5JCM and the camera body is Chameleon CM3-U3-50S5C (Figure 5). Table 4 shows specifications of stereo camera.

Figure 4. Fisheye lens camera: (**a**) Sunnex DSL315 fisheye lens; (**b**) Chameloen3 USB3 Vision.

Table 3. Specifications of fisheye lens camera.

	Lens	Camera Body
Model	Sunnex DSL315	CM3-U3-31S4C
Projection Model	Equisolid angle projection	
Image Size (pixel)	2048 × 1536	
Pixel Size (mm)	0.00345	
Focal Length (mm)	2.67	

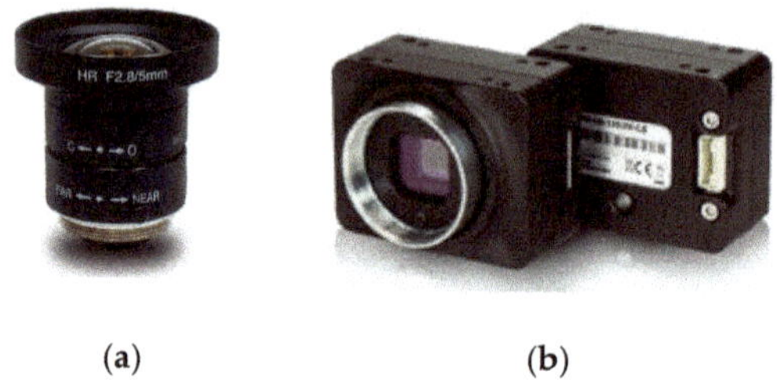

(a) (b)

Figure 5. Stereo camera: (**a**) KOWA LM5JCM; (**b**) Chameleon3 USB3 Vision.

Table 4. Specifications of the stereo camera.

	Lens	Camera Body
Model	KOWA LM5JCM	CM3-U3-50S5C
Projection Model	Perspective	
Image Size (pixel)	2448 × 2048	
Pixel Size (mm)	0.00345	
Focal Length (mm)	5	

3. Mathematical Models

3.1. Point Observation and Systematic Error Model of VLP-16

VLP-16 acquires range and horizontal angle measurements and provides fixed vertical angle for 16 laser channels (described in Section 2.1). The geometric relationship between spherical coordinates (ρ, θ, α) and Cartesian coordinates (x, y, z) is shown in Figure 6. The formulas for converting spherical coordinates (ρ, θ, α) to Cartesian coordinates (x, y, z) are given by Equation (1), where ρ, θ, and α are raw distance measurement, encoder angle measurements, and fixed vertical angle, respectively.

$$P(x,y,z) = \begin{bmatrix} x \\ y \\ z \end{bmatrix} = \begin{bmatrix} \rho\, cos(\alpha) sin(\theta) \\ \rho\, cos(\alpha) cos(\theta) \\ \rho\, sin(\alpha) \end{bmatrix} \tag{1}$$

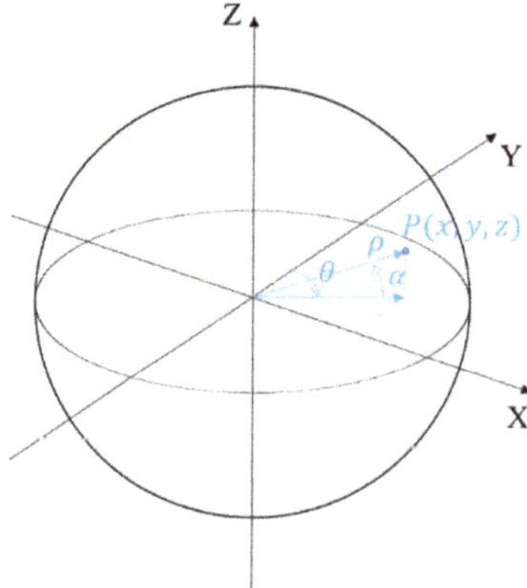

Figure 6. Conversion from a spherical coordinate system to a Cartesian coordinate system.

Sensor modelling is a crucial step in conducting rigorous self-calibration of laser scanners [33,34]. Numerous researchers have independently defined their own TLS error models. The error model of a laser scanner was estimated, and the accuracies of the estimated parameters were determined by comparing the measurements with an electronic distance measurement (EDM) [35]. A later model consists of about 20 additional parameters (APs) and can be found in [36]. This error model effectively modeled TLS instruments; however, it was defined for the AM-CW TLS, which uses a phase-based distance measuring system. In the case of MBL, laser emitter/receiver sets measure distance by a pulse-based Time of Flight (ToF) system with a fixed vertical angle. Therefore, AP terms should be different with phase-based TLS instruments. For Velodyne LiDAR, systematic error coefficients are defined by six parameters (i.e., range offset, scale error, horizontal angular offset, vertical angular offset, horizontal offset, and vertical offset) given by the manufacturer to model the deviations of measurements. Since the number of APs are multiplied by the number of lasers, the AP terms should be carefully chosen to avoid over-parameterization.

Although the manufacturer provides six parameters, some of the APs were neglected in this study. First, the estimation of scale error requires inclusion of an independent scale definition in the self-calibration network [36]. Range scale error cannot be estimated using the calibration approach and must therefore be estimated by other means, and independent baseline testing did not disclose the existence of this error [33]. In performing the tests, therefore, if scale error is included in the adjustment without an independent scale definition, an optimization process is not working properly. Considering there are no a priori known locations of the targets or scanner when performing in situ calibration, the scale error was therefore neglected. Horizontal offset and vertical offset were also fixed to maintain a higher accuracy of adjustment, for the following reasons: (1) Horizontal and vertical offset are highly correlated to the horizontal and vertical rotations, respectively [25,29]. In the case of VLP-16, the correlation coefficients of the parameters corresponding to the vertical and horizontal rotation corrections were found between 0.92 and 0.98, respectively [31]. (2) The vertical and horizontal alignments of each laser are precisely located according to the manufacturer-provided values set below the accuracy of the range observation. (3) Local coordinate error induced by horizontal and vertical offsets is not linearly dependent in the range observation.

Hence, the range offset, horizontal angular offset, and vertical angular offset were considered as APs in this study. Therefore, the point observation model for MBL could finally be determined as Equation (2). The coordinates of ith point at scan position j lying on plane k from laser n are related by rigid body transformation as given by:

$$\begin{bmatrix} X_{ijkn} \\ Y_{ijkn} \\ Z_{ijkn} \end{bmatrix} = R_j \times \begin{bmatrix} \left(\rho_{ijkn} + \Delta\rho_n\right) cos(\alpha_n + \Delta\alpha_n) sin\left(\theta_{ijkn} + \Delta\theta_n\right) \\ \left(\rho_{ijkn} + \Delta\rho_n\right) cos(\alpha_n + \Delta\alpha_n) cos\left(\theta_{ijkn} + \Delta\theta_n\right) \\ \left(\rho_{ijkn} + \Delta\rho_n\right) sin(\alpha_n + \Delta\alpha_n) \end{bmatrix} + \begin{bmatrix} X_j \\ Y_j \\ Z_j \end{bmatrix} \quad (2)$$

where ρ, θ, and α denote a range observation, a horizontal angle observation, and a fixed vertical angle, while $\Delta\rho$, $\Delta\theta$, and $\Delta\alpha$ indicate a range offset, a vertical angular offset, and a horizontal angular offset, respectively. $R_j = R(\kappa_j)R(\phi_j)R(\omega_j)$ is the rotation matrix which transforms the local coordinate system j to the reference coordinate system with the rotation angle ω_j, ϕ_j, and κ_j about the X, Y and Z axes. $\left[X_j\ Y_j\ Z_j\right]^T$ is the translation from j^{th} scan to the reference coordinate system.

3.2. Plane-Based Functional Model

Self-calibration of the laser scanner can be categorized according to the two major point- and plane-based methods. Point-based self-calibration uses center point coordinates extracted from a number of signalized targets through numerous estimation and transformation processes. Point-based self-calibration using TLS such as Trimble GS200 and GX can be found in [37,38]. In addition, research has been studied to determine the optimal network design for correlation mitigation and to achieve good parameterization of TLS self-calibration [39–41]. One limitation of such calibration approaches includes manual installation of signalized targets, which is labor-intensive and can decrease the accuracy of point-based self-calibration due to high parameter correlation [42]. Moreover, in the case of multi-beam laser scanners, extracting the exact target points is almost impossible due to the fixed vertical angle.

Meanwhile, point coordinates on the surface of planar targets can be used directly instead of center point coordinates of signalized targets. Since signalized targets are not required, plane-based self-calibration is one of the most widely adopted methods. The main advantage of plane-based self-calibration is that the plane parameters within each plane can be estimated in the adjustment model, thereby mitigating the need to measure an accurate reference target and enhancing the method's applicability for in situ calibration. Skaloud and Lichti [43] presented a rigorous approach to bore-sight self-calibration of an airborne laser scanning system by conditioning the geo-referenced LiDAR points to fit into common plane surfaces. However, the objective of their work was more oriented to the estimation of extrinsic parameters between the Inertial Measurement Unit (IMU) and the LiDAR unit, considering only range offset as AP. Bae and Lichti [34] conducted plane-based self-calibration with scan data using FARO 880. In their study, self-calibration simulations investigated various scanner configurations, and the results demonstrated that a long baseline between two scan stations enables a more accurate estimation of collimation axis errors. Also, plane-based calibration has been reported to offer almost the same performance as point-based calibration when conducted under a strong network configuration [44]. The self-calibration approach in this study is based on the plane-based functional model of [34]. This model estimates not only exterior orientation parameters (EOPs) and APs, but also plane parameters, simultaneously. The plane-based method can remove the necessity of calibration target set-up and reference target coordinate measurement using additional sensors. The condition equation associated with parameters and observations can be expressed by the plane-based functional model as given by:

$$[a_k \; b_k \; c_k] \times \begin{bmatrix} X_{ijkn} \\ Y_{ijkn} \\ Z_{ijkn} \end{bmatrix} - d_k = 0 \tag{3}$$

where $[a_k \; b_k \; c_k]$ are the direction cosines of the normal vector of the plane k, and d_k is the orthogonal distance from the origin of the reference coordinate system to the plane k. The direction cosines must satisfy the unit length constraint:

$$a_k^2 + b_k^2 + c_k^2 = 1 \tag{4}$$

3.3. *Least Squares Solution*

The combined adjustment model (Gauss–Helmert adjustment model) was used, since the objective function includes inseparable observations and parameters, and each function includes more than one observation. Details on the implementation of the Gauss–Helmert adjustment model can be found in [43]. Therefore, only the quantities of the adjustment will be discussed herein.

First, the VLP-16 provides two observations for each point: a range and a horizontal angle. For unknown parameters, three APs were considered for each laser as aforementioned, and six rigid body transformation parameters for each scan must be included to combine all scanner coordinate systems into a reference coordinate system. Lastly, a unit length condition must be constrained to the equation for each plane. For the network constraint, according to [45], either the ordinary minimum or the inner constraint for the datum definition has no opposing impact on the accuracy of self-calibration. Since the scale is defined by range observations, the ordinary minimum constraint, which fixes the EOPs of the first scan to define the datum as the reference coordinate system, was chosen in this study. In addition, not all 16 laser angular offsets can be estimated simultaneously, because a certain amount of angular offset for every laser can be compensated by sensor orientation, causing a problem when defining the scanner space. Therefore, horizontal and vertical angular offsets for one laser are held fixed. Assuming that we have i points located on k planes from n lasers in j scans in the adjustment, the least squares solution can be summarized as in Table 5.

Table 5. Summary of least squares solution.

Category	Formula
Conditions	$m = i$
Unknowns	$u = 6 \times (j - 1) + (3n - 2) + 4k$
Observations	$l = 2i$
Constraints	$c = k$
Degree of Freedom	$r = m - u + c$

4. Experiment Description and Result Analysis

Calibration experiments were designed using simulated and real datasets. All the experiments were performed by the following process. First, the point cloud for each scan location was captured by defining the "frame", as MBL completes one rotation and covers 360° of the horizontal field of view. Next, plane fitting using Maximum Likelihood Estimation SAmple Consensus (MLESAC) for all point clouds was processed [46]. Each point lying on its surface has a plane number and parameters. Common planes that are mutually detected in all point clouds are manually matched to the reference scan. Least squares adjustment and accuracy assessment follow.

4.1. Simulation Experiments

Since the zero-order design problem—the datum problem—has been addressed by fixing the EOPs of the reference scan as aforementioned, the first-order design problem—the configuration problem—is our interest. Several network configuration conditions were considered to determine the minimum network configuration for plane-based self-calibration of the backpack-based MBL system. These include: (1) the number and configuration of scans; (2) the size, the number, and the configuration of incorporated planes; (3) the minimum number of points lying on the planar surface. In order to determine the minimum network configuration suitable for the backpack-based MBL system in situ self-calibration, simulation experiments were designed with respect to those two significant conditions. Simulation environments were designed by changing the size of the test site and sensor configurations using our own developed simulator program. All the simulated datasets have 0.2° horizontal angle increments, and the same systematic errors. The given systematic errors for the simulation data are shown in Table 6. Random noises were set to 0.003 m, and 0.01° for range and angle observations, respectively.

Table 6. Given systematic error level for simulation experiments.

AP	Values
$\Delta\rho$(m)	0.03
$\Delta\alpha$(°)	0.1
$\Delta\theta$(°)	0.1

The first experiment (Calibration I) was conducted by reducing the number of scans successively to determine the optimal configuration and the minimum number of scans required. As shown in Figure 7a,c, the size of the test site was firstly set to 15 m × 15 m × 3 m, and four scans were located at the corner, two scans were located near the corner, and one was located at the center. This full network configuration, including seven scans and six planes, was constructed based on [45]. Each scan was slightly tilted along the X axes (omega in orientation parameters) for better estimation of the adjustment [42]. Figure 7b,d show the point cloud generated for Calibration I.

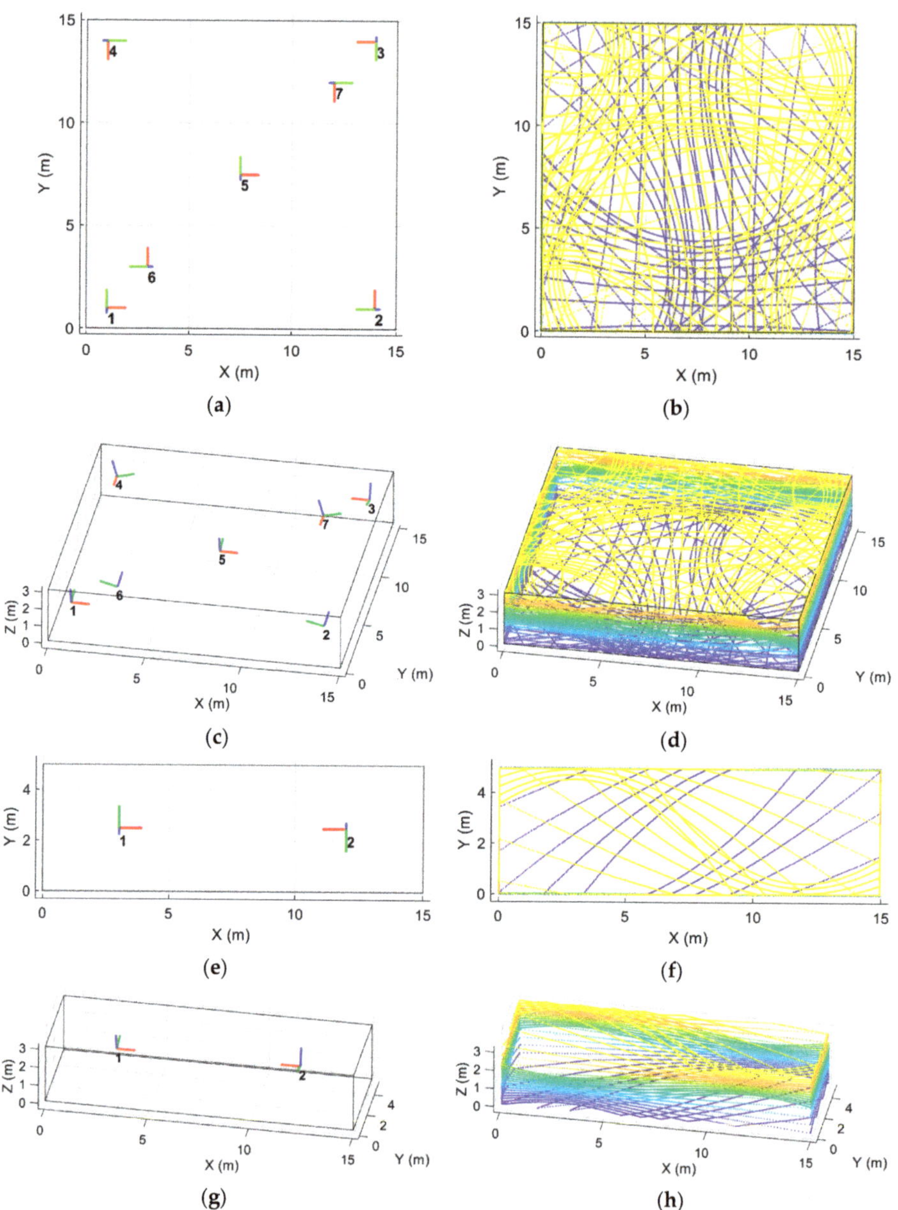

Figure 7. Simulated environment and their corresponding point clouds: (**a**,**c**) network configuration for Calibration I; (**b**,**d**) point cloud generated for Calibration I; (**e**,**g**) network configuration for Calibration II; (**f**,**h**) point cloud generated for Calibration II. Color coded by height.

For the second experiment (Calibration II), as described in Figure 7e,g, the dataset was firstly generated in a 15 m × 5 m × 3 m corridor-shape environment, and the length of the corridor was shortened by 1 m successively until it reached 7 m × 5 m × 3 m in order to investigate the effective dimensions of the room for the self-calibration. After the investigation, the number of incorporated planes also reduces to determine the minimum number of planes required. Figure 7f,h show the point cloud generated for Calibration II. Figure 8 describes the assigned number for each plane.

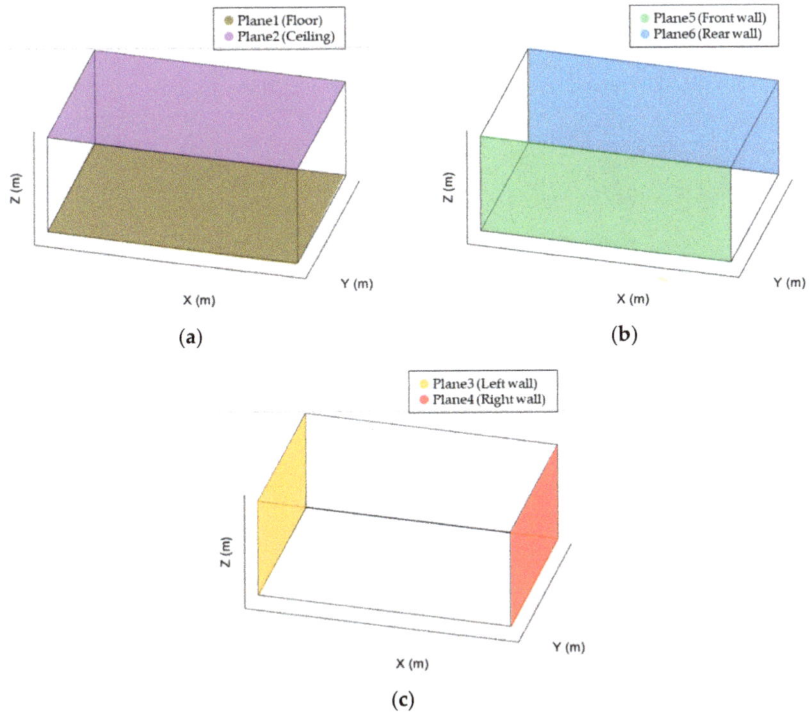

Figure 8. Plane number settings: (**a**) ceiling and floor; (**b**) front and rear walls; (**c**) left and right walls.

For the third experiment (Calibration III), the number of points used in the adjustment reduces to identify the minimum requirement of the redundancy for the adjustment.

4.2. Analysis of Simulation Experiment Results

In the first experiment (Calibration I), the number of scans was reduced successively. Table 7 provides a summary of the first experiment. Also, please refer to Figures 7a–d and 8. The RMSE between estimated and given AP values are plotted in Figure 9. As can be seen, all the tests show high similarity. The asymmetry network configuration of Calibration I-2 might have affected the accuracy of the adjustment. For Calibration I-5, the scan location was too close to the corner, leading to a high incidence angle to the planes. High incidence angle observation tends to deteriorate the overall accuracy of the adjustment [25]. The results from the first experiments indicate that there is no significant change of accuracy when using only two scans compared with the full network, which uses seven scans. For the rest of the experiments, therefore, only two scans (not too close to the corner) were used for self-calibration.

Table 7. Summary of Calibration I (reducing the number of scans).

	Used Scans	Used Planes	Total Points	Used Points	Redundancy
I-1	1, 2, 3, 4, 5, 6, and 7	1, 2, 3, 4, 5, and 6	202,608	2048	1948
I-2	1, 3, 4, 5, 6, and 7		173,664	2048	1954
I-3	1, 3, 5, 6, and 7		144,720	2048	1960
I-4	1, 3, 6, and 7		115,776	2048	1966
I-5	1 and 3		57,888	2048	1978
I-6	6 and 7		57,888	2048	1978

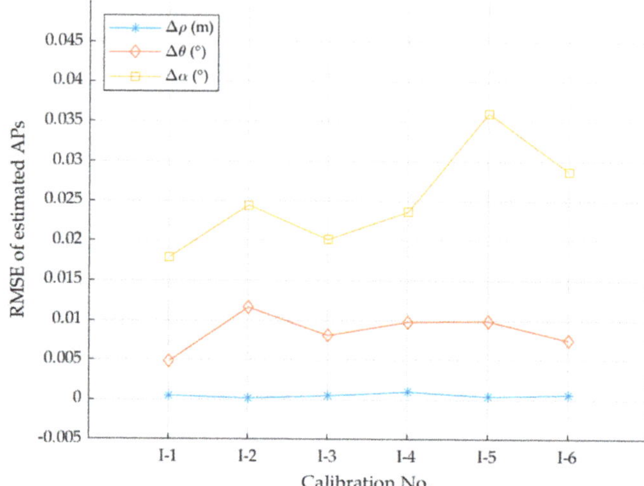

Figure 9. RMSE of estimated additional parameters (APs) for Calibration I.

Following the result from Calibration I, the second experiment (Calibration II) used only two scans for self-calibration by decreasing the dimensions of the room and the number of incorporated planes. Also, please refer to Figures 7e–h and 8. Table 8 summarizes the information on Calibration II. The length of the test site decreased until 7 m. The scan locations were about 3 m apart from the wall and 1 m from each other. Except for Calibration II-13 and II-14, the remaining twelve experiments have calibration solutions. The RMSEs of the estimated APs for the twelve experiments are provided in Figure 10. The range offset showed consistent RMSE values for all experiments, while the two angular offsets showed a slight variance. Even the reduced dimension of the test site (i.e., 7 m × 5 m × 3 m) with three planes (i.e., ceiling and two orthogonal walls) provided a low level of RMSE values as seen in Figure 10.

Table 8. Summary of Calibration II (reducing the length of the corridor and the number of planes).

	Dimensions (m)	Used Planes	Total Points	Used Points	Redundancy	Convergence
II-1	15 × 5 × 3	1, 2, 3, 4, 5, and 6	57,888	2048	1978	O
II-2	14 × 5 × 3					O
II-3	13 × 5 × 3					O
II-4	12 × 5 × 3					O
II-5	11 × 5 × 3					O
II-6	10 × 5 × 3					O
II-7	9 × 5 × 3					O
II-8	8 × 5 × 3					O
II-9	7 × 5 × 3	2, 3, 4, 5, and 6	48,407	2046	1979	O
II-10	7 × 5 × 3	2, 3, 5, and 6	35,446	2048	1984	O
II-11	7 × 5 × 3	2, 3, 4, and 5	38,683	2046	1982	O
II-12	7 × 5 × 3	2, 3, and 5	22,550	2048	1987	O
II-13	7 × 5 × 3	2 and 6	14,194	2048	1990	X
II-14	7 × 5 × 3	2 and 3	11,040	2048	1990	X

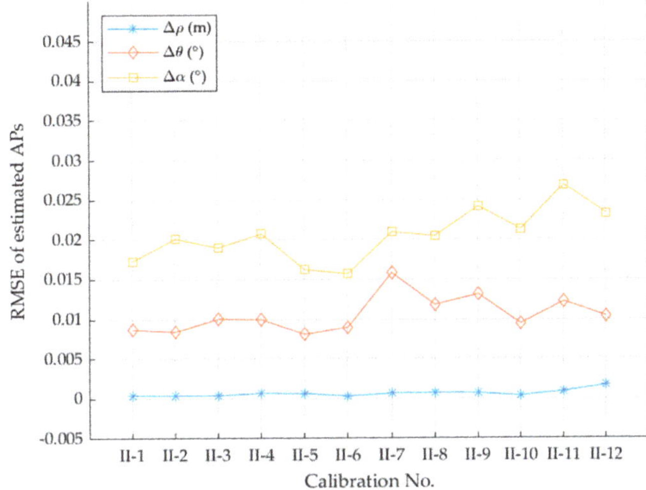

Figure 10. RMSE of estimated APs for Calibration II.

Following the result from Calibration II-12, the third experiment (Calibration III) was conducted by reducing the number of points in the adjustment. In Table 9, the summary of Calibration III is presented. The reduction rate varied from 50 to 99%, which gave the number of used points as 11,275 to 226. The redundancy linearly reduced as the number of points decreased.

Table 9. Summary of Calibration III (Reducing points).

	Used Planes	Total Points	Used Points	Reduction (%)	Redundancy
III-1	2, 3, and 5	22,550	11,275	50	11,214
III-2			9020	60	8959
III-3			6765	70	6704
III-4			4510	80	4449
III-5			2255	90	2194
III-6			1128	95	1067
III-7			902	96	841
III-8			677	97	616
III-9			451	98	390
III-10			226	99	165

As shown in Figure 11, a clear inverse relationship between the RMSE of the estimated parameters and the number of used points was found. We also found that there were dramatic increases in RMSE for angular offsets from Calibration III-9 to III-10, while range offset showed a relatively small increase. To further investigate this phenomenon, additional calibration tests were repeatedly conducted. More specifically, Calibration III-10 (using 226 points) ran five times. At this stage, one should note that the involved number of points was kept as 226 but the points were randomly picked from the whole dataset (i.e., from 22,550 points) for each run. For the comparison, Calibration III-5 (using 2255 points) ran five times as well while picking the involved points randomly for each run. Calibration III-5 was chosen because it corresponded to the inflection point as seen in Figure 11. The results of these additional tests were shown in Figure 12. In the case of repetition of Calibration III-5 (in Figure 12a), RMSE values of the estimated APs for five runs were very similar and the variance of the values was low. On the other hand, repetition of Calibration III-10 (in Figure 12b) showed fluctuating RMSE results, which were too many dataset-dependent outcomes. After these additional tests, it was found that if the number of used points was too small, the calibration process provided unreliable solutions. In this regard, at least 2000 or more points is recommended to estimate reliable angular offsets.

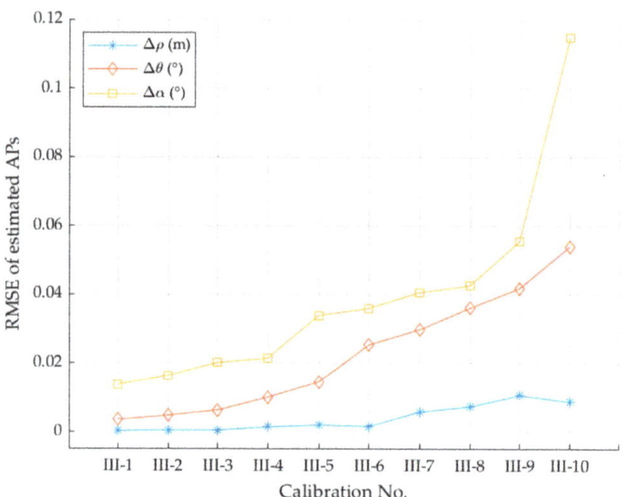

Figure 11. RMSE of estimated APs for Calibration III.

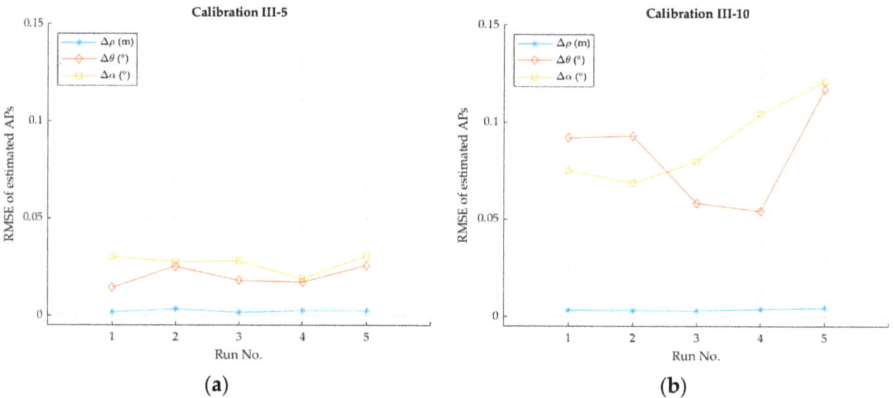

Figure 12. RMSE of estimated Aps: (**a**) five runs of Calibration III-5; (**b**) five runs of Calibration III-10.

4.3. Kinematic Self-Calibration

Based on the results of the simulation tests, a real dataset was captured using the backpack system. The data acquisition site is a corridor of approximate 2.5 m × 40 m × 2.5 m dimensions (Figure 13a). To test the performance of the kinematic in situ self-calibration of the backpack-based MBL system, the user wore the backpack system and walked along the corridor to acquire point clouds. A schematic drawing of the data acquisition site and trajectory are also provided in Figure 13b. As can be seen, two scan locations were selected from whole trajectory. The first scan was captured near the corner, and the second scan was captured apart from the first scan. The omega angle of each scan was slightly tilted, and the kappa angle was rotated by 180°. Then, planar feature extraction and the matching process were carried out using two scan datasets. Figure 14 describes the planar features commonly seen in both scans, which include six vertical planes and two horizontal planes.

A total of eight planes were identified, including walls, doors, a window, the floor, and the ceiling.

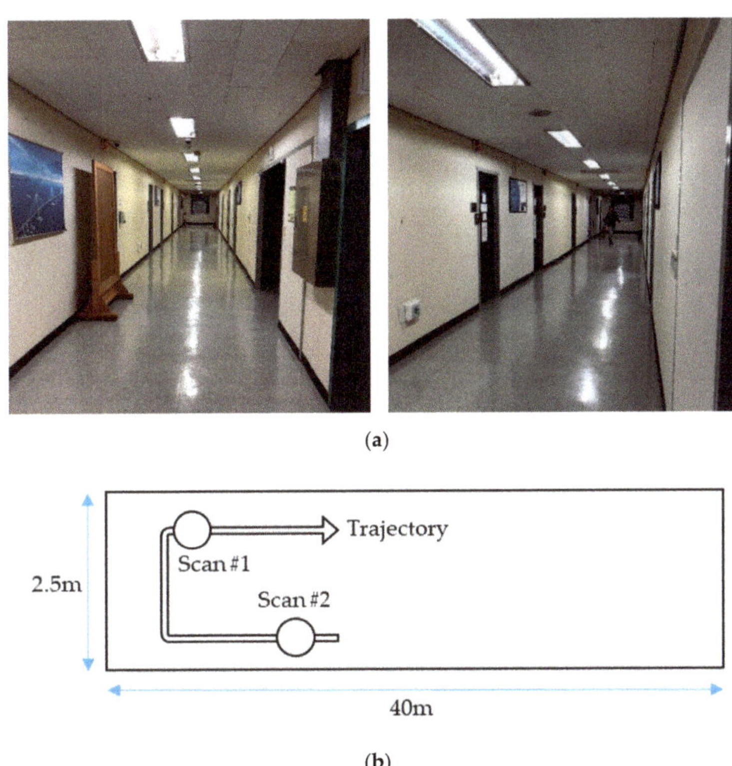

Figure 13. Real dataset acquisition site and schematic set-up. (**a**) Corridor at the Myongji University (**b**) schematic drawing of test site and trajectory.

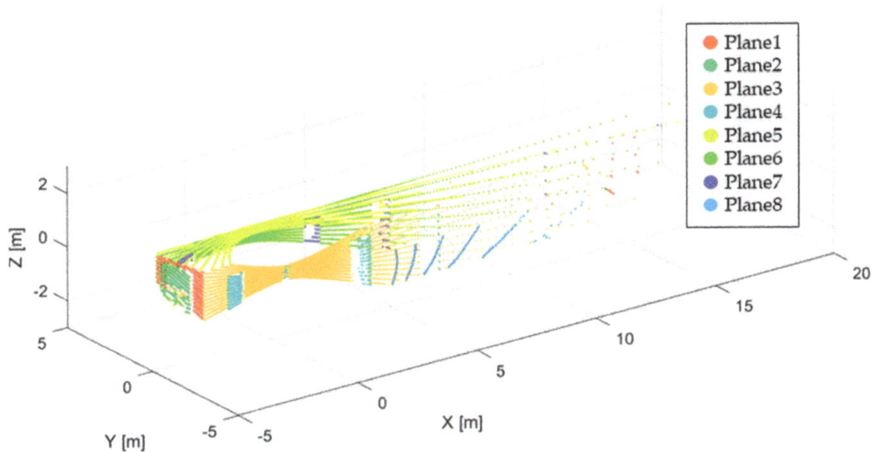

Figure 14. Common planar features seen in Scan #2.

4.4. Analysis of Kinematic Self-Calibration Results

Experiments using the real dataset were conducted to implement the minimum network configuration for the kinematic self-calibration found on the above analysis. Two experiments were conducted to investigate the accuracy when the full and minimum network configurations were used. First, Calibration IV-1 was conducted including all eight planes and 8140 points. Calibration IV-2 was also performed for comparison with Calibration IV-1. Three planes (ceiling and two orthogonal walls) and 2308 points were used for the self-calibration. A summary of the two experiments is presented in Table 10.

Table 10. Summary of Calibration IV.

	Used Planes	Total Points	Used Points	Redundancy
IV-1	1, 2, 3, 4, 5, 6, 7, and 8	54,867	8140	8067
IV-2	1, 3, and 5	41,259	2308	2247

For accuracy evaluation of the kinematic self-calibration, planar misclosure vectors were examined to confirm that the self-calibration had effectively modeled the sensor. For planar misclosure calculations, parameters other than APs were held to the same values in order to compare the results from self-calibration. The planar misclosure results before and after adjustment are given in Figure 15.

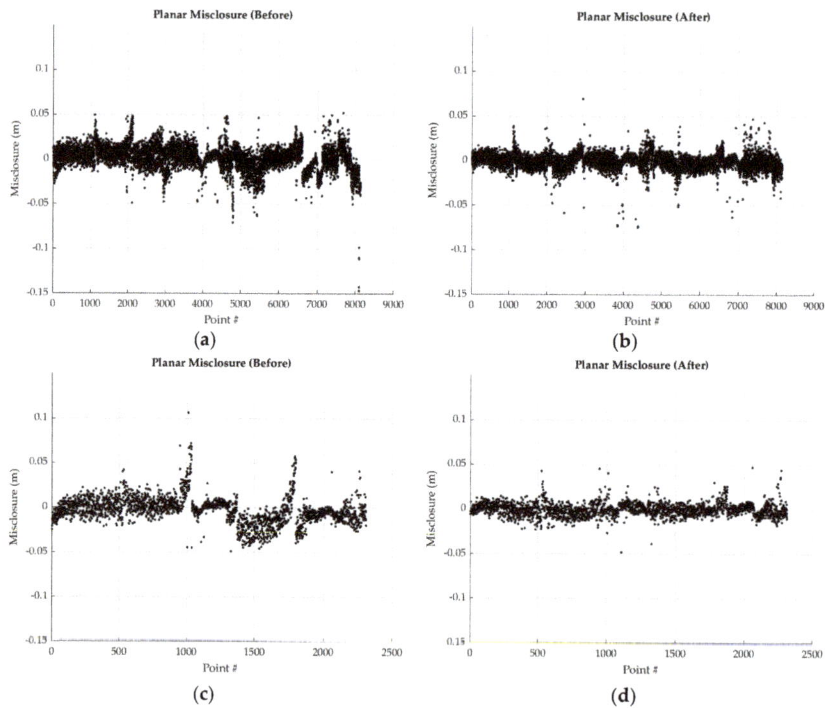

Figure 15. Planar misclosure before and after the adjustment: (**a**) before Calibration IV-1; (**b**) after Calibration IV-1; (**c**) before Calibration IV-2; (**d**) after Calibration IV-2.

As can be seen, in both cases, the systematic errors were not completely removed after the adjustment. Nevertheless, the results of both cases showed improvements of planar misclosure RMSE by 35.7 and 53.3% after the adjustment. The RMSE of Calibration

IV-2 was even lower than that of Calibration IV-1. This is reasonable, as more outliers were included in the Calibration IV-1 dataset. A summary of the results regarding planar misclosure can be found in Table 11.

Table 11. Summary of planar misclosure before and after Calibration IV-1 and IV-2.

		Min (m)	Max (m)	Mean (m)	RMSE (m)	Improvement (%)
IV-1	Before	−0.147	0.052	0.002	0.014	35.7
	After	−0.075	0.070	0.000	0.009	
IV-2	Before	−0.049	0.106	−0.003	0.015	53.3
	After	−0.048	0.047	0.000	0.007	

In order to further investigate the existence of systematic trends, observation residuals from the adjustment were also examined. Figures 16 and 17 describe two observation residuals (a range and a horizontal angle) versus vertical angle, horizontal angle, and range for Calibration IV-1 and IV-2, respectively. For both cases, similar unmodelled systematic errors still existed in the observation residual. Residuals versus vertical angle showed no trends of systematic effects, as the mean residual values with respect to laser elevation angle had zero values (refer to Figure 16a,b and Figure 17a,b). On the other hand, residuals versus horizontal angle showed large variations at 90° and 270°, which are the directions of high incidence angles to the walls (refer to Figure 16c,d and Figure 17c,d). This was expected from simulation tests and previous studies in the literature. For the final analysis for residuals, residuals versus range were plotted (refer to Figure 16e,f and Figure 17e,f). As can be seen, outliers increased as range increased, in both the range and horizontal angle observations.

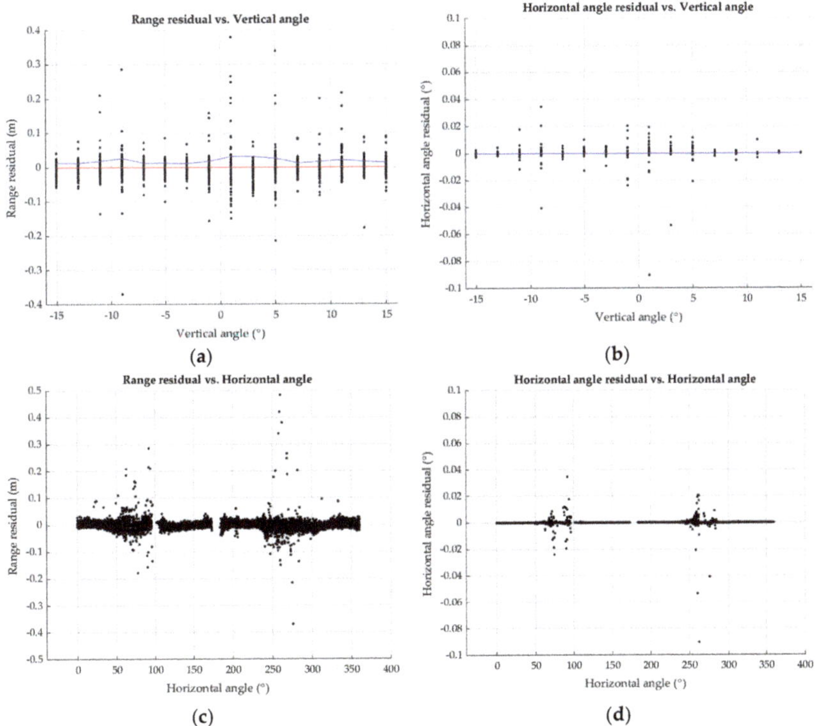

Figure 16. *Cont.*

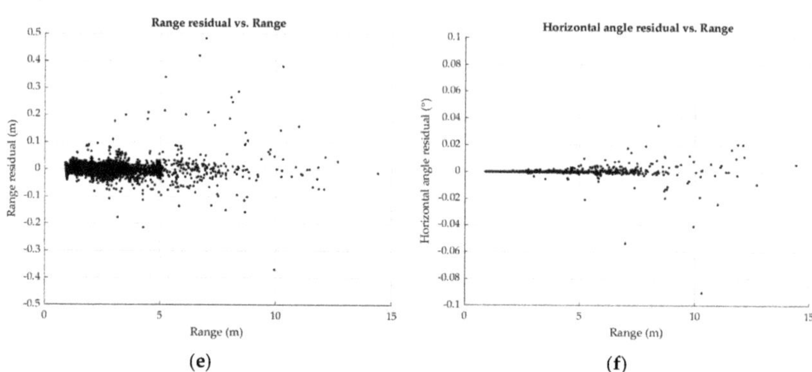

(e) (f)

Figure 16. Measurement residuals of Calibration IV-1: (**a**,**c**,**e**) range residuals; (**b**,**d**,**f**) horizontal angle residuals; red line and blue line in (**a**,**b**) mean average and RMSE of residuals, respectively.

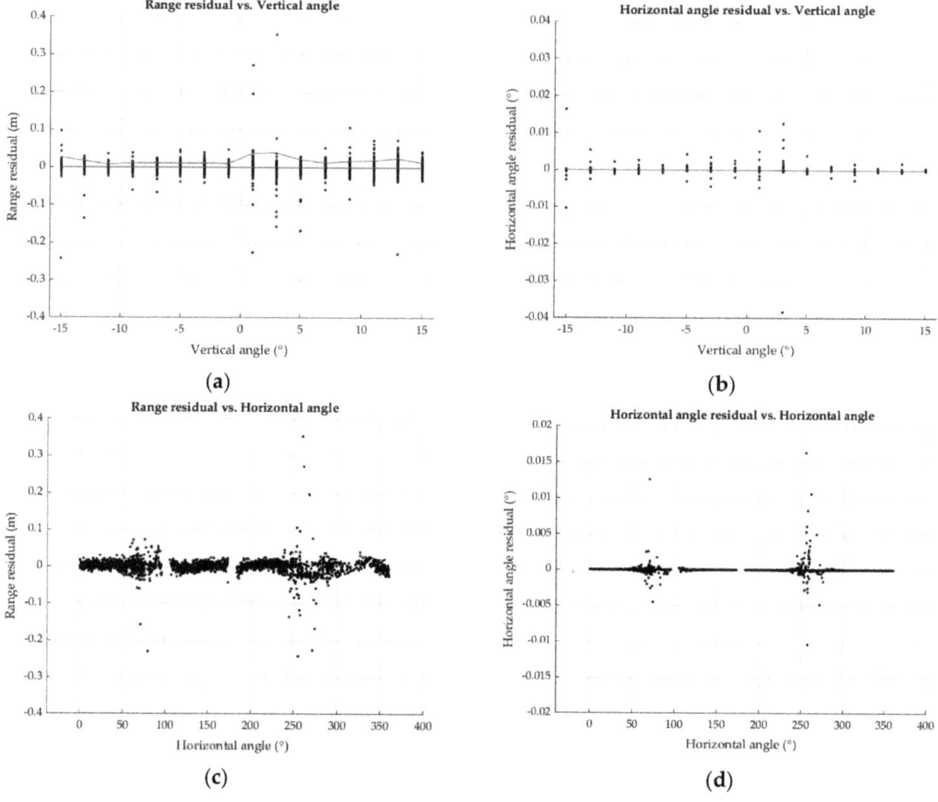

(a) (b)

(c) (d)

Figure 17. *Cont.*

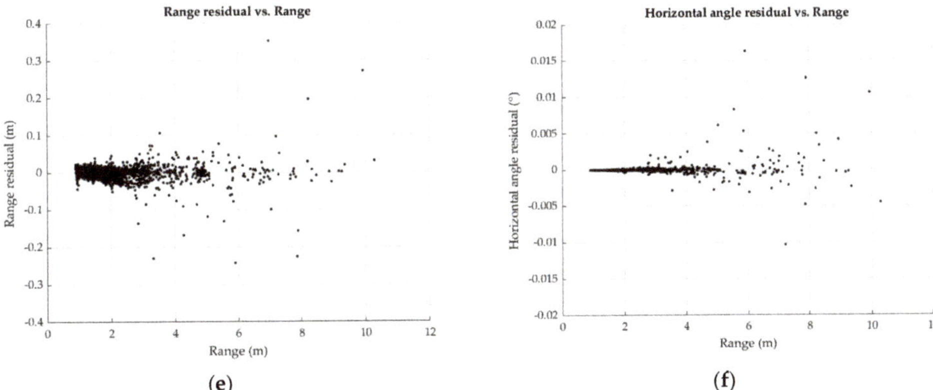

Figure 17. Measurement residuals of Calibration IV-2: (**a**,**c**,**e**) range residuals; (**b**,**d**,**f**) horizontal angle residuals; red line and blue line in (**a**,**b**) mean average and RMSE of residuals, respectively.

Furthermore, the reliability of the estimated parameter could be presumed by examining the correlation coefficients between the estimated parameters. The correlation coefficients between the APs and EOPs are presented in Tables 12 and 13. The correlations between the EOPs and APs were averaged over all of the lasers with one EOP, and the correlations between APs were averaged over all lasers with the same laser. In general, correlations between the EOPs were higher than the correlations between the EOPs and the APs. The moderate (not significantly strong) correlation values were, also, found between Z_o and ω, Z_o and ϕ, and Y_o and κ. They were marked in bold in the tables. This correlation between Z_o and the orientation parameters was likely due to the small number of points with respect to the horizontal plane (floor and ceiling), and, moreover, the amount of tilting angle was not sufficient for perfectly de-coupling between the translation and orientation parameters. Nevertheless, it should be noted that the estimated parameters did not have strong correlations and were derived reliably.

Table 12. Averaged correlation coefficients between exterior orientation parameters (EOPs) and APs for Calibration IV-1.

	X_o	Y_o	Z_o	ω	ϕ	κ	$\Delta\rho$	$\Delta\theta$	$\Delta\alpha$
X_o		0.077	0.039	0.050	0.074	0.041	0.045	0.030	0.012
Y_o			0.056	0.099	0.083	**0.666**	0.086	0.284	0.023
Z_o				**0.458**	**0.761**	0.080	0.034	0.070	0.139
ω					0.067	0.277	0.048	0.059	0.183
ϕ						0.070	0.093	0.086	0.148
κ							0.032	0.191	0.138
$\Delta\rho$								0.080	0.046
$\Delta\theta$									0.070

Based on the results from Calibration IV-1 and IV-2, the estimated AP values and their standard deviations for all the lasers are plotted in the function of the vertical angle in Figure 18. Both calibrations demonstrated similar results of parameter estimation, while the estimation of parameters with respect to the directed high vertical angle laser showed different results. As can be seen, the standard deviations of horizontal angular offset for directing a high vertical angle laser were high, while the standard deviations of the vertical angular offset were low. The horizontal and vertical angular offsets for laser 1 (having 1° of vertical angle) were held fixed as zeros, as mentioned in Section 3.3. Estimated parameters for Calibrations IV-1 and IV-2 are provided in Table 14. The EOP estimation showed reasonable standard deviations relative to the estimated values. As similarly investigated in Section 4.2, kinematic calibration of the backpack-based MBL system can be performed using a minimum network configuration with reasonable accuracy in a real environment.

Table 13. Averaged correlation coefficients between EOPs and APs for Calibration IV-2.

	X_o	Y_o	Z_o	ω	ϕ	κ	$\Delta\rho$	$\Delta\theta$	$\Delta\alpha$
X_o		0.032	0.099	0.041	0.049	0.098	0.064	0.061	0.015
Y_o			0.033	0.183	0.024	**0.675**	0.151	0.270	0.019
Z_o				**0.553**	**0.721**	0.096	0.084	0.054	0.128
ω					0.146	0.088	0.038	0.057	0.148
ϕ						0.081	0.151	0.085	0.130
κ							0.175	0.154	0.123
$\Delta\rho$								0.093	0.097
$\Delta\theta$									0.082

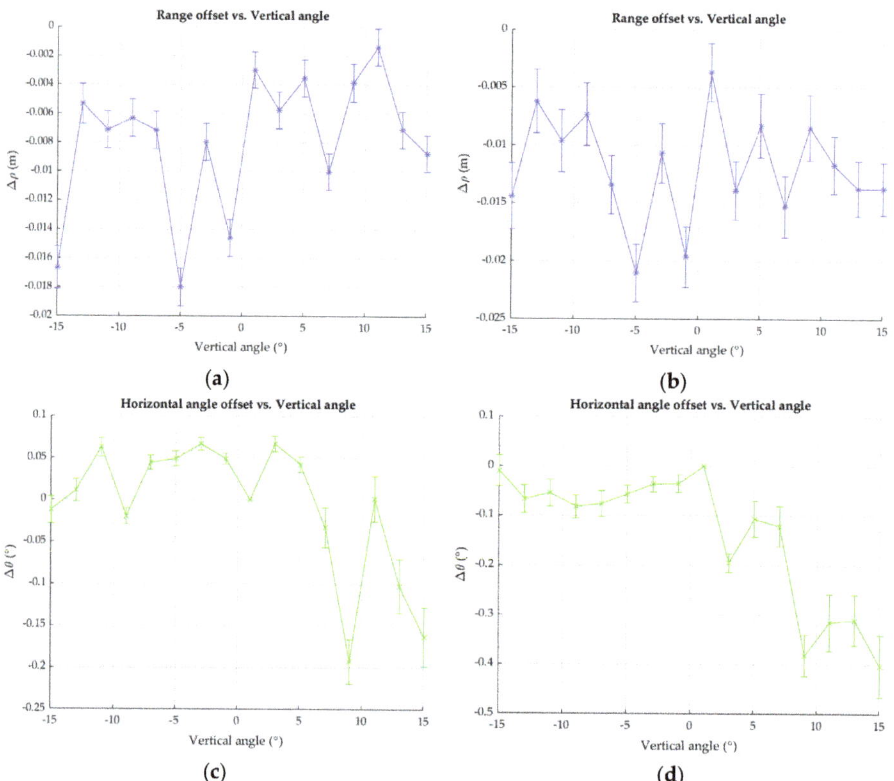

Figure 18. *Cont.*

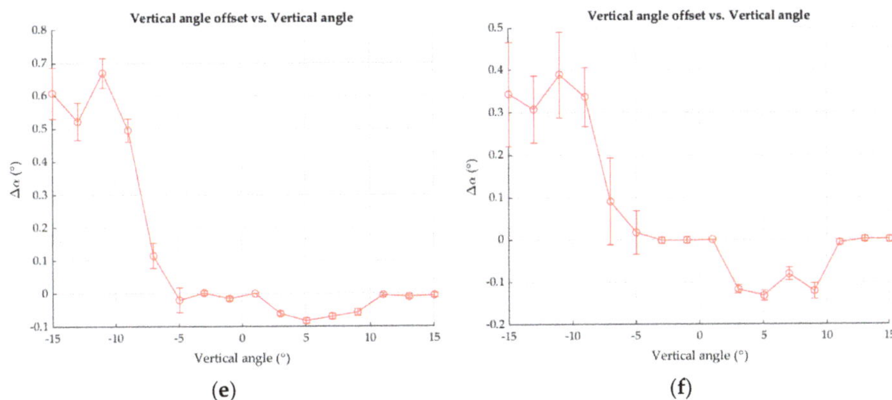

Figure 18. Estimated AP values and their standard deviations versus vertical angle: (**a**,**c**,**e**) Calibration IV-1; (**b**,**d**,**f**) Calibration IV-2.

Table 14. Estimated parameters and their standard deviation for Calibration IV-1 and IV-2.

	Estimated Values					
	Calibration IV-1			Calibration IV-2		
$X_o(m)$		2.876 ± 0.002			2.870 ± 0.003	
$Y_o(m)$		0.689 ± 0.001			0.686 ± 0.002	
$Z_o(m)$		0.105 ± 0.001			0.101 ± 0.001	
$\omega(°)$		−0.356 ± 0.031			−0.359 ± 0.055	
$\phi(°)$		−0.131 ± 0.011			−0.133 ± 0.019	
$\kappa(°)$		−3.078 ± 0.008			−3.076 ± 0.017	
Laser No.	**Δρ(m)**	**Δθ(°)**	**Δα(°)**	**Δρ(m)**	**Δθ(°)**	**Δα(°)**
0	−0.017 ± 0.001	−0.012 ± 0.016	0.608 ± 0.078	−0.014 ± 0.003	−0.010 ± 0.032	0.344 ± 0.123
1	−0.003 ± 0.001	-	-	−0.004 ± 0.003	-	-
2	−0.005 ± 0.001	0.011 ± 0.013	0.522 ± 0.057	−0.006 ± 0.003	−0.068 ± 0.028	0.308 ± 0.079
3	−0.006 ±0.001	0.067 ± 0.009	−0.061 ± 0.007	−0.014 ± 0.003	−0.195 ± 0.019	−0.118 ± 0.010
4	−0.007 ± 0.001	0.063 ± 0.011	0.668 ± 0.046	−0.010 ± 0.003	−0.055 ± 0.027	0.389 ± 0.101
5	−0.004 ± 0.001	0.042 ± 0.009	−0.082 ± 0.008	−0.008 ± 0.003	−0.107 ± 0.036	−0.133 ± 0.012
6	−0.006 ± 0.001	−0.02 ± 0.010	0.496 ± 0.034	−0.007 ± 0.003	−0.082 ± 0.023	0.337 ± 0.070
7	−0.010 ± 0.001	−0.033 ± 0.024	−0.069 ± 0.008	−0.015 ± 0.003	−0.122 ± 0.040	−0.082 ± 0.015
8	−0.007 ± 0.001	0.044 ± 0.009	0.114 ± 0.038	−0.013 ± 0.003	−0.077 ± 0.026	0.090 ± 0.102
9	−0.004 ± 0.001	−0.193 ± 0.026	−0.057 ± 0.010	−0.008 ± 0.003	−0.382 ± 0.042	−0.121 ± 0.018
10	−0.018 ± 0.001	0.049 ± 0.009	−0.02 ± 0.037	−0.021 ± 0.002	−0.057 ± 0.018	0.017 ± 0.051
11	−0.001 ± 0.001	0.001 ± 0.027	−0.004 ± 0.006	−0.012 ± 0.002	−0.316 ± 0.058	−0.008 ± 0.007
12	−0.008 ± 0.001	0.067 ± 0.008	0.003 ± 0.007	−0.011 ± 0.003	−0.037 ± 0.016	−0.002 ± 0.007
13	−0.007 ± 0.001	−0.103 ± 0.032	−0.01 ± 0.006	−0.014 ± 0.002	−0.311 ± 0.052	0.000 ± 0.007
14	−0.015 ± 0.001	0.049 ± 0.007	−0.015 ± 0.007	−0.020 ± 0.003	−0.035 ± 0.018	−0.002 ± 0.007
15	−0.009 ± 0.001	−0.163 ± 0.035	−0.006 ± 0.006	−0.014 ± 0.002	−0.403 ± 0.062	0.000 ± 0.007

4.5. Temporal Stability Analysis of Kinematic Self-Calibration

An additional experiment (Calibration V) was conducted for temporal stability analysis. About one month later after performing self-calibration (Calibration IV), we recalibrated the same sensor under the similar condition. More specifically, the data acquisition site and the rest of the conditions were identical to the Calibration IV for comparison. Based on the findings from the previous analysis, the minimum network configuration was considered for Calibration V. Summary of Calibration V is given in Table 15.

Table 15. Summary of Calibration V.

	Used Planes	Total Points	Used Points	Redundancy
V	1, 3, and 5	23,049	2305	2244

Table 16 shows the summary of planar misclosure before and after Calibration V. The planar misclosure results before and after Calibration V are, also, given in Figure 19. As can be seen, the RMSE of planar misclosure before the calibration (i.e., 0.038 m) increased compared to Calibration IV-2 (i.e., 0.015 m). In addition, estimated AP values from Calibration V can be found in Figure 20, showing different values compared to Calibration IV-2 (Figure 18b,d,f). On the other hand, RMSE of planar misclosure after Calibration V (i.e., 0.007 m in Table 16) showed a similar result to Calibration IV-2 (i.e., 0.007 m in Table 11). Through this additional experiment, we found that the APs of the MBL system were changing unstably over time, and the calibration process provided a stable level of planar misclosure RMSE.

Table 16. Summary of planar misclosure before and after Calibration V.

		Min (m)	Max (m)	Mean (m)	RMSE (m)	Improvement (%)
V	Before	−0.223	0.124	0.011	0.038	81.6
	After	−0.025	0.046	0.000	0.007	

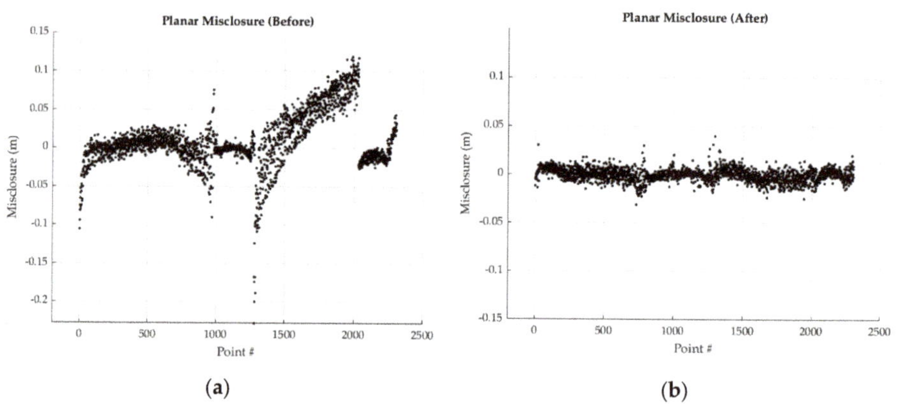

Figure 19. Planar misclosure before and after the adjustment: (**a**) before Calibration V; (**b**) after Calibration V.

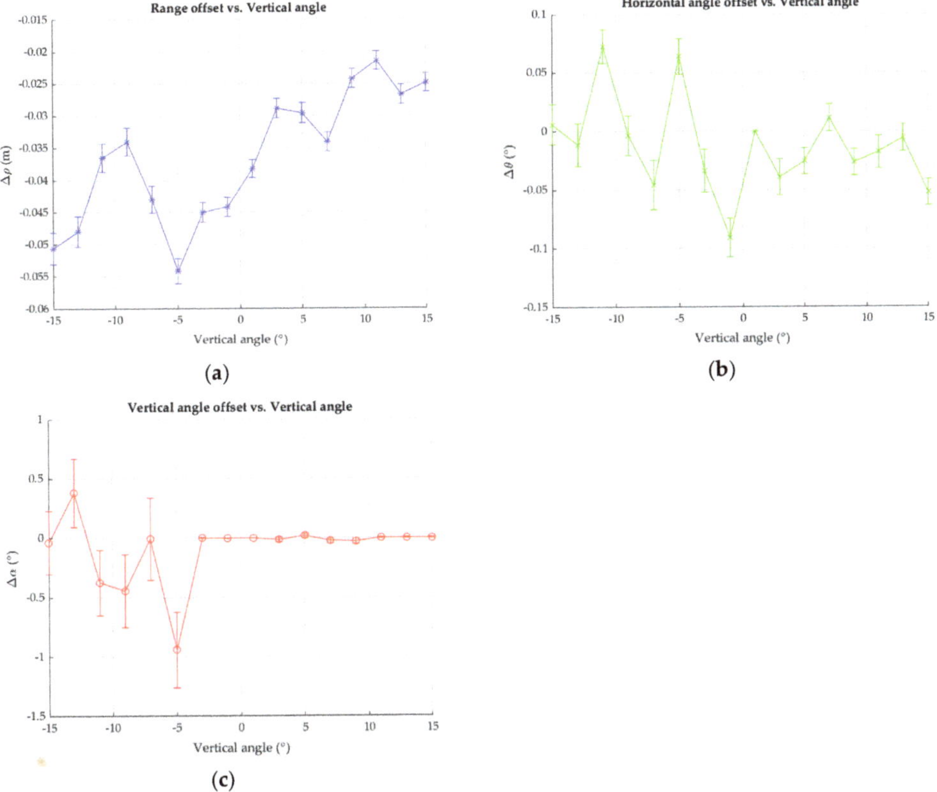

Figure 20. Estimated AP values and their standard deviations versus vertical angle for Calibration V: (**a**) range offset; (**b**) horizontal angle offset; (**c**) vertical angle offset.

5. Discussion

In this section, the effectiveness of the proposed self-calibration method is discussed in terms of accuracy, time, and cost. First, a comparative analysis between the previous MBL self-calibration approaches and the proposed one is carried out. Table 17 shows the performances (i.e., accuracy and improvement) of five representative MBL calibration approaches and the proposed one. At this stage, one should note that a direct comparison of the performances among different approaches is almost impossible since their sensor systems, methods, datasets, environments, computational performances are all different. Nevertheless, the comparison in Table 17 shows the overall performances of the approaches. Although there is a variance in the type of MBL sensors, our method achieved a fair improvement (35–81%) compared to other results. In particular, Glennie [29] and Chan and Lichti [30] performed kinematic calibrations of MBL mounted on a vehicle platform. In these approaches, the RMSEs of planar misclosure after the calibration were 0.023 and 0.014 m, respectively, which are higher than the proposed case. This is due to the fact that vehicle moving speed is much faster than human walking speed, causing a higher measurement noise level. The proposed approach showed a similar (or somewhat better) level of RMSE to the static calibrations.

Table 17. Comparison with other self-calibration results.

Approaches	RMSE of Planar Misclosure		Improvement	Sensor
	Static	Kinematic		
Glennie and Lichti [25]	0.013 m		63.8%	HDL-64E
Chen and Chien [28]	0.013 m		40%	HDL-64E
Glennie [29]		0.023 m	37.8%	HDL-64E
Chan and Lichti [30]	0.008 m	0.014 m	41–71%	HDL-32E
Glennie et al. [31]	0.025 m		20%	VLP-16
Proposed		0.007 m	35–81%	VLP-16

Secondly, the proposed kinematic self-calibration of the backpack-based MBL system can significantly reduce time and cost compared to the traditional target-based static calibrations. The proposed one does not need an installation of targets and tripods to fix the scan location since it is aiming for kinematic in situ calibration. Although the proposed method is not fully automatic, the running time of the whole process takes up to 30 s (under the condition of 2 scans, 3 planes, and around 2000 points), except for the manual plane matching. The computer processor is an AMD Ryzen 5 1600 six-core processor with DDR-4 16GB 1500MHz of RAMs. The program for the whole process is written in MATLAB. The program and the algorithm are not yet optimized, and the processing time can be improved in the future study.

6. Conclusions

This study investigated kinematic in situ self-calibration to frequently re-calibrate the backpack-based MBL system using on-site data for handling unstable measurements of the sensor. In order to determine the minimum network configuration for kinematic self-calibration, simulation experiments were conducted beforehand. First, a full network of the simulated datasets was generated, and self-calibrations were performed by reducing the number of scans, the size of the test site, the number of incorporated planes, and the number of points. The accuracies of the experiments were analyzed by examining the RMSE of the estimated APs to determine the minimum network configuration. The results of the simulation experiments show relatively stable performance with a minimum network configuration of at least two scans, three planes that are orthogonal to each other, and around two thousand points used. Based on this preliminary analysis, kinematic self-calibration using real datasets was then performed. The datasets were acquired while the user was wearing a backpack system and walking along a corridor. The accuracy of kinematic self-calibration was evaluated by investigating the planar misclosure, measurement residuals, correlation coefficients, and estimated parameters and their standard deviations. The results demonstrate that the kinematic self-calibration of the backpack-based MBL system could improve the point cloud accuracy with the RMSE of planar misclosure up to 81%. Moreover, the effectiveness of the proposed approach in terms of time and cost was also addressed.

After the various experiments and analysis using the proposed kinematic in situ self-calibration of the backpack-based MBL system, the contributions of this study can be summarized as follows. First, self-calibration of MBL was analyzed with respect to various network configurations. The minimum network configuration for the kinematic in situ self-calibration of the backpack-based MBL system and its performance were found through various experiments. Secondly, the kinematic in situ self-calibration of the backpack-based MBL system can perform using on-site datasets, reaching the higher accuracy of point cloud. In addition, this research can serve as a guideline for users who require the self-calibration of a backpack-based MBL system to improve overall accuracy and to generate point cloud data for precise mapping or surveying. Future studies will mostly focus on: (1) the development of real-time automatic plane matching for automatic in situ calibration; (2) outlier removal during the iteration of least squares based on statistical analysis to maximize calibration accuracy.

Author Contributions: H.S.K. was responsible for developing the methodology and writing the original manuscript, C.K. provided the backpack system, Y.K. helped revise the manuscript, and K.H.C. supervised the research. All authors have read and agreed to the published version of the manuscript.

Funding: This work was supported by a National Research Foundation of Korea (NRF) grant (no. 2019R1A2C1011014) funded by the Korean government (MSIT).

Informed Consent Statement: Not applicable.

Data Availability Statement: Data sharing not applicable.

Acknowledgments: The authors are grateful to C2L Equipment (www.c2l-equipment.com) for helping to develop the backpack system for this research.

Conflicts of Interest: The authors declare no conflict of interest.

References

1. Bosché, F.; Ahmed, M.; Turkan, Y.; Haas, C.T.; Haas, R. The value of integrating scan-to-BIM and scan-vs-BIM techniques for construction monitoring using laser scanning and BIM: The case of cylindrical MEP components. *Autom. Construct.* **2015**, *49*, 201–213. [CrossRef]
2. Cole, D.M.; Newman, P.M. Using laser range data for 3D SLAM in outdoor environments. In Proceedings of the 2006 IEEE International Conference on Robotics and Automation (ICRA), Orlando, FL, USA, 15–19 May 2016; pp. 1556–1563.
3. Wu, Y.; Kim, H.; Kim, C.; Han, S.H. Object recognition in construction-site images using 3D CAD-based filtering. *J. Comput. Civ. Eng.* **2009**, *24*, 56–64. [CrossRef]
4. Shih, N.J.; Huang, S.T. 3D scan information management system for construction management. *J. Const. Eng. Manag.* **2006**, *132*, 134–142. [CrossRef]
5. Tang, P.; Anil, E.B.; Akinci, B.; Huber, D. Efficient and Effective Quality Assessment of As-Is Building Information Models and 3D Laser-Scanned Data. In Proceedings of the International Workshop on Computing in Civil Engineering 2011, Miami, FL, USA, 12–22 June 2011; pp. 486–493.
6. Park, H.S.; Lee, H.M.; Adeli, H.; Lee, I. A new approach for health monitoring of structures: Terrestrial laser scanning. *Comp. Aid. Civ. Inf. Eng.* **2007**, *22*, 19–30. [CrossRef]
7. Yang, H.; Xu, X.; Neumann, I. Laser scanning-based updating of a finite-element model for structural health monitoring. *IEEE Sens. J.* **2015**, *16*, 2100–2104. [CrossRef]
8. Riveiro, B.; Lindenbergh, R. *Laser Scanning: An Emerging Technology in Structural Engineering*; CRC Press: Boca Raton, FL, USA, 2019; pp. 1–3.
9. Halterman, R.; Bruch, M. Velodyne HDL-64E lidar for unmanned surface vehicle obstacle detection. In Proceedings of the Unmanned Systems Technology XII, Orlando, FL, USA, 6–9 April 2010; Volume 7692, p. 76920D.
10. Moosmann, F.; Stiller, C. Velodyne SLAM. In Proceedings of the IEEE Intelligent Vehicles Symposium 2011, Baden-Baden, Germany, 5–9 June 2011; pp. 393–398.
11. Geiger, A.; Lenz, P.; Urtasun, R. Are we ready for autonomous driving? The kitti vision benchmark suite. In Proceedings of the 2012 IEEE Conference on Computer Vision and Pattern Recognition (CVPR), Providence, RI, USA, 16–21 June 2012; pp. 3354–3361.
12. Choi, K.H.; Kim, Y.; Kim, C. Analysis of Fish-Eye Lens Camera Self-Calibration. *Sensors* **2019**, *19*, 1218. [CrossRef]
13. Jozkow, G.; Toth, C.; Grejner-Brzezinska, D. Uas Topographic Mapping with Velodyne Lidar Sensor. *ISPRS Ann. Photogramm. Remote Sens. Spat. Inf. Sci.* **2016**, *III-1*, 201–208. [CrossRef]
14. Chen, T.; Dai, B.; Liu, D.; Song, J.; Liu, Z. Velodyne-based curb detection up to 50 m away. In Proceedings of the 2015 IEEE Intelligent Vehicles Symposium, Seoul, Korea, 28 June–1 July 2015; pp. 241–248.
15. Hess, W.; Kohler, D.; Rapp, H.; Andor, D. Real-time loop closure in 2D LIDAR SLAM. In Proceedings of the 2016 IEEE International Conference on Robotics and Automation (ICRA), Stockholm, Sweden, 16–21 May 2016; pp. 1271–1278.
16. Ravi, R.; Lin, Y.J.; Elbahnasawy, M.; Shamseldin, T.; Habib, A. Bias impact analysis and calibration of terrestrial mobile lidar system with several spinning multibeam laser scanners. *IEEE Trans. Geosci. Remote Sens.* **2018**, *56*, 5261–5275. [CrossRef]
17. Shamseldin, T.; Manerikar, A.; Elbahnasawy, M.; Habib, A. SLAM-based Pseudo-GNSS/INS Localization System for Indoor LiDAR Mobile Mapping Systems. In Proceedings of the IEEE/OIN PLANS 2018, Monterey, CA, USA, 23–26 April 2018.
18. Leica Pegasus: Backpack Wearable Mobile Mapping Solution. Available online: https://leica-geosystems.com/products/mobile-sensor-platforms/capture-platforms/leica-pegasus-backpack# (accessed on 29 December 2020).
19. Viametris. Available online: https://www.viametris.com/backpackmobilescannerbms3d (accessed on 29 December 2020).
20. GreenValley International. Available online: https://greenvalleyintl.com/hardware/libackpack/ (accessed on 29 December 2020).
21. Gexcel Geomatics & Excellence. Available online: https://gexcel.it/en/solutions/heron-mobile-mapping (accessed on 31 December 2020).

22. Velas, M.; Spanel, M.; Sleziak, T.; Habrovec, J.; Herout, A. Indoor and Outdoor Backpack Mapping with Calibrated Pair of Velodyne LiDARs. *Sensors* **2019**, *19*, 3944. [CrossRef]
23. Chow, J.C. Multi-Sensor Integration for Indoor 3D Reconstruction. Ph.D. Thesis, University of Calgary, Calgary, AB, Canada, April 2014.
24. García-San-Miguel, D.; Lerma, J.L. Geometric calibration of a terrestrial laser scanner with local additional parameters: An automatic strategy. *ISPRS J. Photogramm. Remote Sens.* **2013**, *79*, 122–136. [CrossRef]
25. Glennie, C.; Lichti, D.D. Static calibration and analysis of the velodyne HDL-64E S2 for high accuracy mobile scanning. *Remote Sens.* **2010**, *2*, 1610–1624. [CrossRef]
26. Muhammad, N.; Lacroix, S. Calibration of a Rotating Multi-Beam Lidar. In Proceedings of the IEEE /RSJ International Conference on Intelligent Robots and Systems (IROS), Toulouse, France, 18–22 October 2010; pp. 5648–5653.
27. Atanacio-Jiménez, G.; González-Barbosa, J.-J.; Hurtado-Ramos, J.B.; Francisco, J.; Jiménez-Hernández, H.; García-Ramirez, T.; González-Barbosa, R. Velodyne HDL-64E calibration using pattern planes. *Int. J. Adv. Robot. Syst.* **2011**, *8*, 70–82. [CrossRef]
28. Chen, C.-Y.; Chien, H.-J. On-site sensor recalibration of a spinning multi-beam LiDAR system using automatically-detected planar targets. *Sensors* **2012**, *12*, 13736–13752. [CrossRef]
29. Glennie, C. Calibration and kinematic analysis of the Velodyne HDL-64E S2 Lidar sensor. *Photogramm. Eng. Remote Sens.* **2012**, *78*, 339–347. [CrossRef]
30. Chan, T.O.; Lichti, D.D. Automatic In Situ Calibration of a Spinning Beam LiDAR System in Static and Kinematic Modes. *Remote Sens.* **2015**, *7*, 10480–10500. [CrossRef]
31. Glennie, C.L.; Kusari, A.; Facchin, A. Calibration and stability analysis of the VLP-16 laser scanner. *Int. Arch. Photogramm. Remote Sens. Spat. Inf. Sci.* **2016**, *40*, 55–60. [CrossRef]
32. Glennie, C.; Lichti, D. Temporal stability of the Velodyne HDL-64E S2 scanner for high accuracy scanning applications. *Remote Sens.* **2011**, *3*, 539–553. [CrossRef]
33. Lichti, D.; Licht, M.G. Experiences with terrestrial laserscanner modelling and accuracy assessment. *Int. Arch. Photogramm. Remote Sens.* **2006**, *36*, 155–160.
34. Bae, K.H.; Lichti, D. On-site self-calibration using planar features for terrestrial laser scanners. In Proceedings of the ISPRS Workshop on Laser Scanning and SilviLaser, Espoo, Finland, 12–14 September 2007.
35. Lichti, D.D.; Stewart, M.P.; Tsakiri, M.; Snow, A.J. Calibration and Testing of a Terrestrial Laser Scanner. In Proceedings of the International Archives of Photogrammetry and Remote Sensing, Amsterdam, The Netherlands, 16–22 July 2000; Volume XXXIII. Part B5.
36. Lichti, D. Error modelling, calibration and analysis of an AM-CW terrestrial laser scanner system. *ISPRS J. Photogramm. Remote Sens.* **2007**, *61*, 307–324. [CrossRef]
37. Chow, J.; Ebeling, A.; Teskey, W. Low cost artificial planar target measurement techniques for terrestrial laser scanning. In Proceedings of the FIG Congress 2010: Facing the Challenges—Building the Capacity, Sydney, Australia, 11–16 April 2010.
38. Chow, J.C.; Lichti, D.D.; Teskey, W.F. Self-calibration of the Trimble (Mensi) GS 200 terrestrial laser scanner. *Int. Arch. Photogramm. Remote Sens. Spat. Inf. Sci.* **2010**, *38*, 161–166.
39. Lichti, D.D. A review of geometric models and self-calibration methods for terrestrial laser scanners. *Bol. Cienc. Geod.* **2010**, *16*, 3–19.
40. Lichti, D. Terrestrial laser scanner self-calibration: Correlation sources and their mitigation. *ISPRS J. Photogramm. Remote Sens.* **2010**, *65*, 93–102. [CrossRef]
41. Reshetyuk, Y. Self-Calibration and Direct Georeferencing in Terrestrial Laser Scanners. Ph.D. Thesis, Royal Institute of Technology, Stockholm, Sweden, 2009.
42. Chow, J.; Lichti, D.; Glennie, C.; Hartzell, P. Improvements to and comparison of static terrestrial LiDAR self-calibration methods. *Sensors* **2013**, *13*, 7224–7249. [CrossRef] [PubMed]
43. Skaloud, J.; Lichti, D. Rigorous approach to bore-sight self-calibration in airborne laser scanning. *ISPRS J. Photogramm. Remote Sens.* **2006**, *61*, 47–59. [CrossRef]
44. Chow, J.; Lichti, D.; Teskey, W. Accuracy assessment of the Faro Focus3D and Leica HDS6100 panoramic type terrestrial laser scanner through point-based and plane-based user self-calibration. In Proceedings of the FIG Working Week 2012: Knowing to Manage the Territory, Protect the Environment, Evaluate the Cultural Heritage, Rome, Italy, 6–10 May 2012.
45. Abbas, M.A.; Lichti, D.D.; Chong, A.K.; Setan, H.; Majid, Z. An on-site approach for the self-calibration of terrestrial laser scanner. *Measurement* **2014**, *52*, 111–123. [CrossRef]
46. Torr, P.H.; Zisserman, A. MLESAC: A new robust estimator with application to estimating image geometry. *Comput. Vis. Image Und.* **2000**, *78*, 138–156. [CrossRef]

Article

A Learning-Based Image Fusion for High-Resolution SAR and Panchromatic Imagery

Dae Kyo Seo [1] and Yang Dam Eo [2,*]

[1] Department of Advanced Technology Fusion, Konkuk University, Seoul 05029, Korea; tjeory@konkuk.ac.kr
[2] Department of Civil & Environmental Engineering, Konkuk University, Seoul 05029, Korea
* Correspondence: eoandrew@konkuk.ac.kr; Tel.: +82-450-3078

Received: 31 March 2020; Accepted: 7 May 2020; Published: 9 May 2020

Abstract: Image fusion is an effective complementary method to obtain information from multi-source data. In particular, the fusion of synthetic aperture radar (SAR) and panchromatic images contributes to the better visual perception of objects and compensates for spatial information. However, conventional fusion methods fail to address the differences in imaging mechanism and, therefore, they cannot fully consider all information. Thus, this paper proposes a novel fusion method that both considers the differences in imaging mechanisms and sufficiently provides spatial information. The proposed method is learning-based; it first selects data to be used for learning. Then, to reduce the complexity, classification is performed on the stacked image, and the learning is performed independently for each class. Subsequently, to consider sufficient information, various features are extracted from the SAR image. Learning is performed based on the model's ability to establish non-linear relationships, minimizing the differences in imaging mechanisms. It uses a representative non-linear regression model, random forest regression. Finally, the performance of the proposed method is evaluated by comparison with conventional methods. The experimental results show that the proposed method is superior in terms of visual and quantitative aspects, thus verifying its applicability.

Keywords: image fusion; random forest regression; SAR image; panchromatic image; high-resolution

---

1. Introduction

Recently, various high-resolution satellite sensors have increasingly been developed, especially the synthetic aperture radar (SAR) imaging sensor, which has an important advantage in Earth observations [1,2]. It is an active sensor that provides its own source of illumination, which is independent of solar illumination and is not affected by daylight or night darkness [3]. It can also penetrate through atmospheric effects, allowing for Earth observation regardless of weather conditions such as rain, fog, smoke, or clouds [4,5]. Information contained in a SAR image depends on the backscattering characteristics of the surface targets and is sensitive to the geometry of the targets [6]. The image provides information on surface roughness, object shape, orientation, or moisture content [7,8]. Furthermore, the SAR image can highlight objects that have a low contrast in optical images. However, interpreting the details in SAR images is a challenging task for several reasons: (1) SAR images inherently contain geometric distortions due to distance-dependence along the range axis and signatures related to radar signal wavelengths [9]; (2) the human eye is familiar with the visible part of the electromagnetic spectrum and is not adapted to the microwave-scattering phenomenon [10]; (3) the reflectance properties of objects in the microwave range depend on the frequency band used and may significantly differ from the usual assumption of diffuse reflection at the Earth's surface [11]; (4) since SAR images are inherently coherent during the process of their generation, speckle noise is inevitable in the resulting images, rendering the images unintuitive [12]; and (5) such images also contain the after-effects caused by foreshortening, slant-range scale distortion, layover, and shadows [13,14]. Thus, the SAR image can be visually difficult to interpret and, ultimately, this data improvement approach is designed at the end to be implemented in the monitoring and analyzing earth surface issues that offering an advanced solution for many applications including environmental studies [15].

To improve the quality and interpretability of SAR images, image fusion with optical images, which contain information regarding reflective and emissive characteristics, can be a good alternative [16–18]. In particular, the panchromatic image can be utilized because it is physically sensitive to ground objects and reflects the objects' contour information with high spatial resolution and abundant textural features [19]. The overall concept of image fusion between the SAR and panchromatic images is to incorporate spatial details extracted from the panchromatic image into the SAR image by using an appropriate algorithm [20]. Therefore, the fusion of the SAR and panchromatic images makes it possible to use complementary information and contributes to a better understanding of the objects in target areas [21]. Furthermore, the fusion of SAR and panchromatic images has additional benefits, such as the sharpening of image quality, enhancement of certain features that are invisible with either data set in the non-combined state, complementation of data sets for improved classification, detection of changes using multi-temporal data, and substitution of missing information in one image with signals from another sensor image [1].

However, because of the significant differences between the imaging mechanisms of the SAR and optical sensors, the generation of surface features of the same object are different in SAR and panchromatic images [5]. Conventional image-fusion methods such as principal component analysis (PCA) and high-pass filtering are not appropriate because they do not consider the differences in imaging mechanisms and the spectrum characteristics between the two image types [22]. An alternative approach is multiscale decomposition, based on which various methods have been proposed for the fusion of SAR and panchromatic images; however, these methods have some limitations [19,20,22,23]. For image fusion based on these methods, the SAR and panchromatic images are represented by the fixed orthogonal basis function, and the image fusion is performed by the fusion of the coefficients of different sub-bands in the transform domain [24]. Because some features cannot be represented sparsely, this fusion cannot represent all useful features accurately due to limited fixed transforms [20]. In particular, the discrete wavelet transforms (DWT) fusion method only uses features of single pixels to make decisions, and it is not shift-invariant [25]. Similarly, the contourlet transform (CT)-based fusion method lacks shift-invariance, which results in pseudo-Gibbs phenomena around singularities, and it has difficulty in preserving edge information. The non-subsampled contourlet transform (NSCT)-based method, which is a fully shift-invariant form of the CT, leads to better frequency selectivity and regularity [26]. However, this method still fails to fuse the features of physically heterogenous images [5]. Another approach is the sparse representation method, in which the generation of dictionary and sparse coding is crucial [24]. This method can extract potential information from input images in addition to representing them sparsely; however, this method does have limitations. Firstly, the advanced sparse coefficients fusion rule may cause spatial inconsistency, and secondly, the trained dictionary cannot accurately reflect the complex structure and detail of the input images [27].

To overcome these limitations, this study proposes a new image-fusion method that utilizes useful features as much as possible and considers the differences in imaging mechanisms. Instead of directly fusing pixels or decomposing them to perform fusion in a limited transform, this algorithm aims to extract sufficient features and establish relationships to fuse the SAR and panchromatic images. This makes it possible to contain the structural and detailed information of panchromatic images and increase the overall interpretability of SAR images [28]. Furthermore, a learning-based approach is used to account for the differences in imaging mechanisms of the SAR and panchromatic images. Random forest (RF) regression, which can model non-linear relationships, is utilized, and learning is performed for each class to reduce the complexity of the algorithm and for better predictions [29,30]. Then, experiments are performed on multiple scenes to demonstrate the capability and performance of the proposed method. The results are comprehensively compared with those of conventional image-fusion methods. The main contributions of this study can be summarized as follows: (1) this is the first learning-based approach for fusing single high-resolution SAR and panchromatic images; (2) to consider the differences in imaging mechanisms, this method uses RF regression, which can model non-linear patterns, avoids overfitting, and is relatively robust to the presence of noise; (3) this method performs classification of the image, where the complexity is reduced by establishing relationships for each class.; and (4) this method extracts various features to consider sufficient information.

The rest of this paper is organized as follows: Section 2 describes materials used in detail and the proposed algorithm in detail. In Section 3, the results of the proposed method are presented, and they are compared with those of the conventional image-fusion methods and discussed. Finally, Section 4 concludes the paper.

2. Materials and Methods

2.1. Study Site and Dataset

The study areas are Gwangjin-gu and Seongdong-gu, located in Seoul, in central South Korea (Figure 1). These areas are covered by forests, grass, barren land, water, and developed structures; thus, they represent an extensive range of terrain morphologies. The dataset used in the experiments for the panchromatic image type is WorldView-3, and for the SAR image type, the Korea Multi-Purpose Satellite-5 (KOMPSAT-5) dataset is used. The WorldView-3 dataset used in this study was acquired on 7 August 2015; the images in this dataset have a spatial resolution of 0.31 m and a radiometric resolution of 11 bits [31]. The KOMPSAT-5 dataset was acquired on 10 September 2015; it was obtained in the enhanced high-resolution mode with a spatial resolution of 1 m, an ascending orbit, and horizontal transmit-horizontal receive (HH) polarization. The processing level was L1D, which performs terrain correction and then geolocates onto a digital elevation model (DEM) with cartographic projection [32]. Initially, speckle noise exists in the SAR images; however, it is expected to reduce through filtering, thereby providing better information. In this study, a gamma map filter of 5×5 kernels, which is the most efficient filter for reducing speckles while preserving object edges, is selected for speckle filtering [33]. Furthermore, because the filtered KOMPSAT-5 images should be calculated with the same weights as the weights used for the WordlView-3 images, the filtered KOMPSAT-5 images are configured with a matching pixel value range [20]. For the fusion scheme, the KOMPSAT-5 images are resampled at a resolution of 0.31 m to match that of the WorldView-3 images. Next, to remove the misregistration error term, image registration is applied using manual ground control points, followed by geometric transformation. In addition, the coordinate system of each image is projected as the Universal Transverse Mercator Coordinate System (UTM). Then, for a reasonable computation time, experiments are performed with subsets of 2000×2000 pixels, and the total area of the three sites is selected to validate the proposed method. Table 1 describes the specifications of the data, and Figures 2–4 show the preprocessed experimental images for three sites.

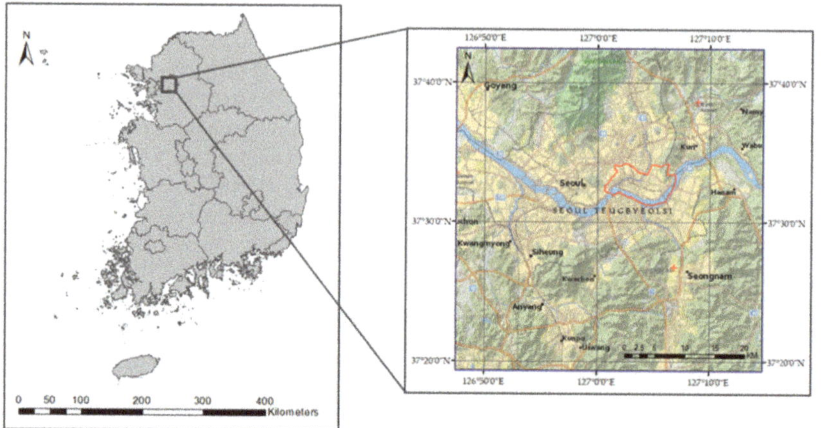

Figure 1. Location of the study area (red outline indicates the study area).

Table 1. Specifications of the experimental data (SAR: synthetic aperture radar, HH: horizontal transmit-horizontal receive).

Sensor	KOMPSAT-5 (SAR Image)	WorldView-3 (Panchromatic Image)
Location		Seoul (Korea)
Date	10 September 2015	7 August 2015
Spatial resolution (m)	1 m	0.31 m
Radiometric resolution	-	11-bit
Polarization	HH	-
Flight direction	Ascending	-
Image size (pixels)		2000 × 2000

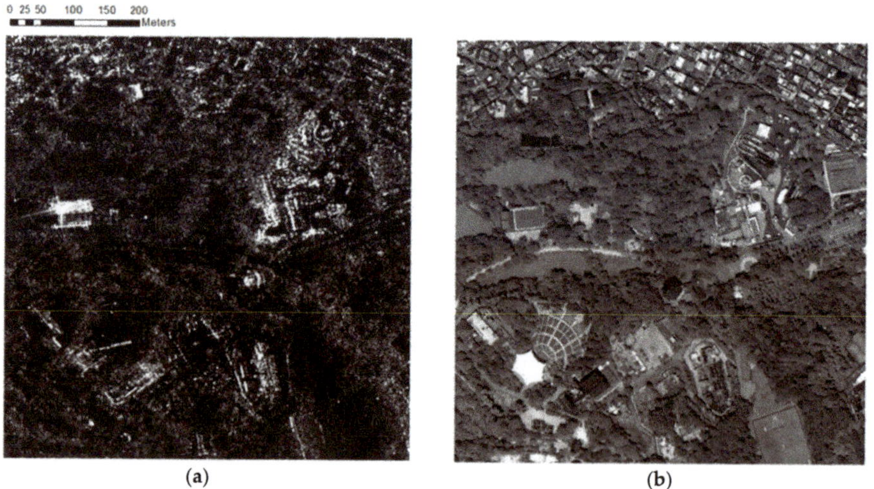

Figure 2. Experimental images from Site 1: (**a**) synthetic aperture radar image acquired on 10 September 2015; and (**b**) panchromatic image acquired on 7 August 2015.

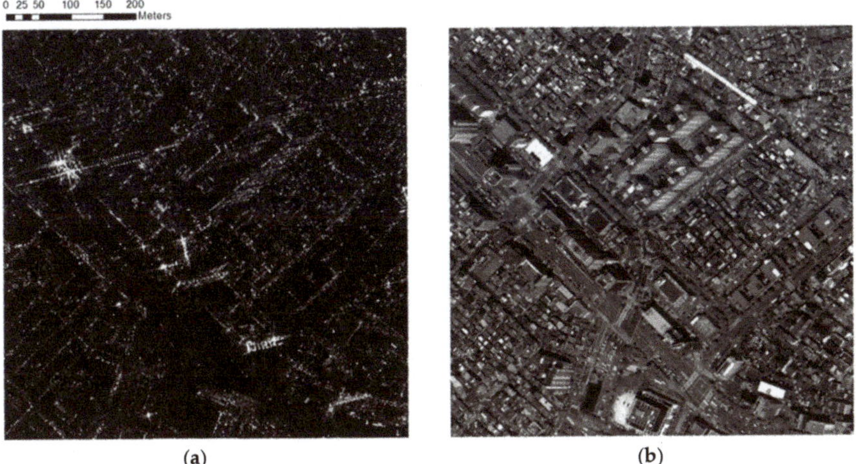

Figure 3. Experimental images from Site 2: (**a**) synthetic aperture radar image acquired on 10 September 2015; and (**b**) panchromatic image acquired on 7 August 2015.

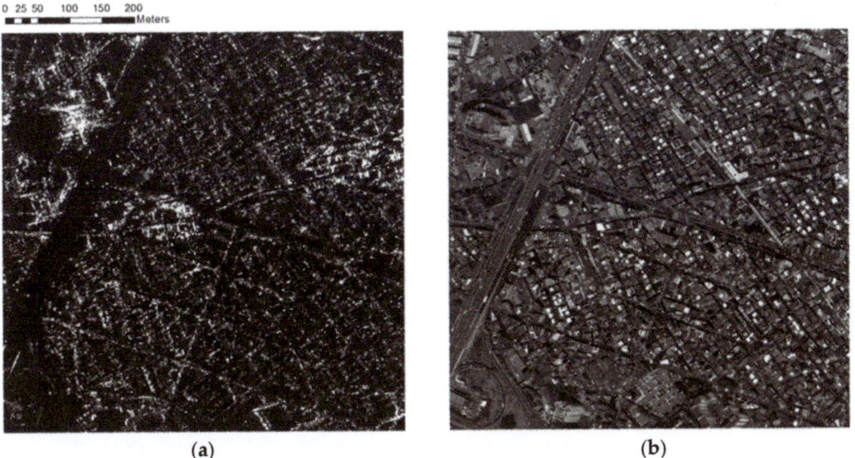

Figure 4. Experimental images from Site 3: (**a**) synthetic aperture radar image acquired on 10 September 2015; and (**b**) panchromatic image acquired on 7 August 2015.

2.2. Methods

The proposed fusion framework can be decomposed into four steps for the preprocessed images: (1) selection of training pixels, (2) classification, (3) feature extraction, and (4) learning-based image fusion; they are shown in Figure 5. In the first step, the pixels to be used for the training are selected, and in the second step, classification is performed on the SAR and panchromatic images. In the third step, feature descriptors are extracted to be used for training as the SAR image, and in the fourth step, fusion is performed by establishing a relationship based on learning. These steps are described below.

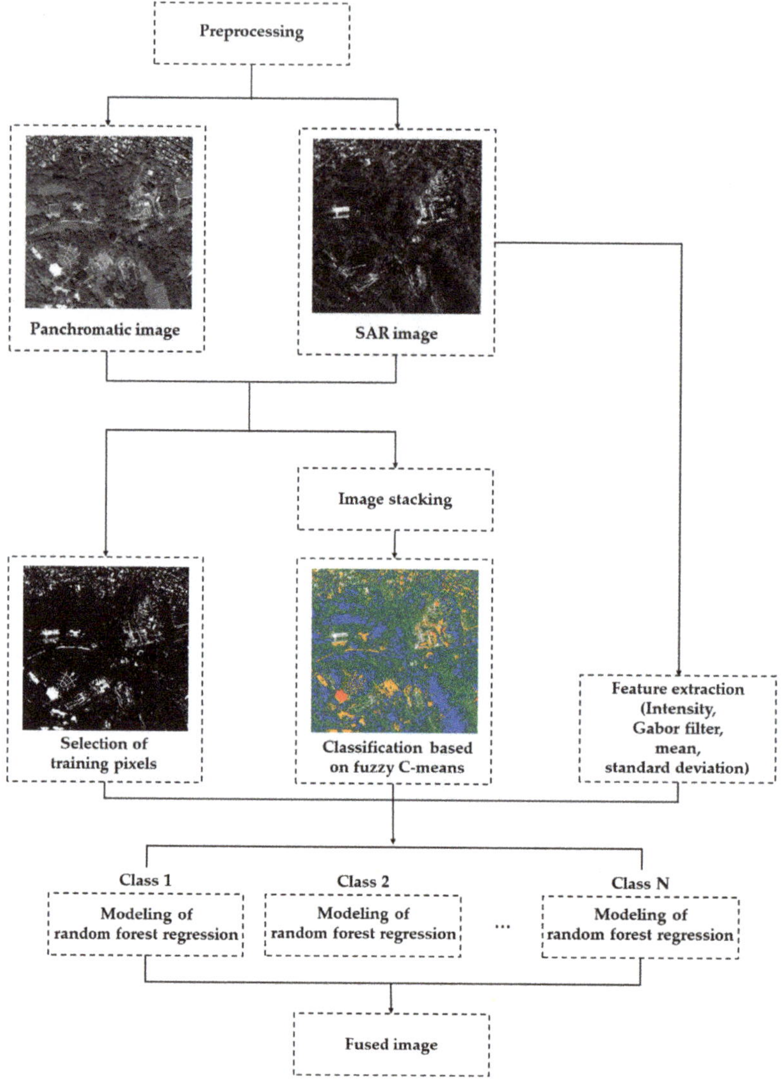

Figure 5. Flowchart of the proposed method.

2.2.1. Selection of Training Pixels

In the step involving the selection of training pixels, meaningful pixels to be used for establishing the relationship should be selected. In particular, training pixels should be selected to consider the differences in imaging mechanisms. This study selects invariant pixels, that is, pixels with little difference in reflectance between the two images. In other words, the relationships are established based on invariant pixels, and the values of pixels with substantially large differences are predicted [28,34]. The invariant pixels are acquired through image differencing, which is a method that subtracts pixel values between the SAR and panchromatic images, in accordance with Equation (1):

$$\Delta x_d(i,\ j) = I_S(i,j) - I_P(i,j) + C \tag{1}$$

where I_S is the pixel value in the SAR image, I_P is the pixel value in the panchromatic image, i and j represent rows and columns, respectively, and C is an arbitrary constant. Then, Otsu's method is used to classify change and no-change, where the no-change region is selected as the invariant pixels.

2.2.2. Classification

To reduce the complexity of the algorithm and to enforce a higher prediction, classification is performed in this study [28]. In other words, each class is obtained, and learning is performed independently for each class. Here, classification is performed by stacking two images to consider the characteristics of both SAR and panchromatic images using fuzzy C-means (FCM), which is an unsupervised classification algorithm [35]. FCM is based on the optimization of the objective function based on the similarity measure considering the distance between data and the center of the cluster as shown by Equation (2):

$$J(U,V) = \sum_{n=1}^{N}\sum_{c=1}^{C} u_{kn}{}^{m} d^2(y_n,\ v_k) \tag{2}$$

where N is the number of data; c is the number of clusters; u_{kn} is the membership function and satisfied the condition $0 \le u_{kn} \le 1$, $\sum_{k=1}^{c} u_{kn} = 1$; m is a weighting exponent that control the degree of fuzziness in the resulting membership functions and is set to 2 for simplicity [36]; $d^2(y_n,\ v_k) = \|y_n - v_k\|^2$ is squared distance between y_n and v_k, in which $Y = [y_1,\ y_2,\ \cdots y_n]$ is a dataset to be grouped and v_k is the cluster center. To minimize the objective function, the FCM algorithm performs an iterative process, and the membership functions and cluster centers are defined as Equations (3) and (4):

$$u_{kn} = \frac{1}{\sum_{j=1}^{c}\left(\frac{d^2(y_n,\ v_k)}{d^2(y_n,\ v_j)}\right)^{\frac{1}{(m-1)}}} \tag{3}$$

$$v_k = \frac{\sum_{n=1}^{N} u_{kn}{}^{m} y_n}{\sum_{n=1}^{N} u_{kn}{}^{m}} \tag{4}$$

U and V are iteratively updated to obtain an optimal solution, and the iterative process ends when $\|U^{(r)} - U^{(r-1)}\| < \varepsilon$, where $U^{(r)}$ and $U^{(r-1)}$ are the membership functions in the rth and $r-1$th iterations and ε is a predefined small positive threshold [37]. Furthermore, the number of clusters is a key parameter in the proposed method as it determines the number of training models in which land-cover distribution characteristics as well as performance and training time should be considered. If there are not enough clusters, the land-cover distribution characteristics will be neglected, and if there are too many clusters, the training time will increase, complex computations will be necessary, and overtraining may occur. Therefore, in this study, the number of clusters is set to 6 to not only obtain appropriate performance and training times but also to consider the land cover distribution characteristics [28].

2.2.3. Feature Extraction

Conventional image-fusion methods use only the pixel values of SAR and panchromatic images. However, in general, the gray level of single pixels is not informative; therefore, additional information other than the pixel values is necessary [38,39]. To ensure that abundant information is considered, this study uses texture information. Several approaches exist for extracting texture features, for example, the gray-level co-occurrence matrix, local binary patterns, and Gabor filters [40–42], among which the Gabor filter is selected; this filter is inspired by a multi-channel filtering theory for processing visual information in the human visual system [43]. It is advantageous in terms of invariance to illumination, rotation, scale, and translation; thus, it has been successfully applied for various image processing and machine vision applications [44]. The 2-D Gabor function comprises a complex sinusoid modulated by a Gaussian envelope, in which the Gabor filter includes a real component and an imaginary one. In this study, because of the substantial magnitude of the images, only real components were considered. Calculation without imaginary components would cause small discrepancies; however, the results are still efficient in terms of feature extraction time [45]. This can be represented as Equations (5)–(7):

$$G(a,b) = \exp\left(-\frac{a'^2 + \gamma^2 b'^2}{2\sigma^2}\right)\cos\left(2\pi\frac{a}{\lambda} + \varphi\right) \quad (5)$$

$$a' = a\cos\theta + b\sin\theta \quad (6)$$

$$b' = -a\sin\theta + b\cos\theta \quad (7)$$

where a and b are pixel positions; γ is the spatial aspect ratio (the default value is 0.5 in [46]); σ is the standard deviation of the Gaussian envelope; λ is the wavelength of the sinusoidal factor and $1/\lambda$ is equal to the spatial frequency f; φ is the phase offset, where $\varphi = 0$ and $\varphi = \pi/2$ return the real component and imaginary component, respectively [47]; and θ is the orientation.

Gabor features, generally taken as Gabor filters, are constructed by selecting different spatial frequencies and orientations. The frequency corresponds to scale information and is expressed as Equation (8):

$$f_m = k^{-m} f_{max};\ m = \{0,\ 1,\ \cdots M-1\} \quad (8)$$

where f_m is the mth frequency; f_{max} is the central frequency of the filter at the highest frequency, for which the most commonly adopted value is $\sqrt{2}/4$, based on the suggestion that f_{max} should be lower than 0.5 [48,49]; and k is the scale factor, which this study selects as 2 [50]. Then, the orientations are expressed as Equation (9):

$$\theta_n = \frac{n2\pi}{N};n = \{0,\ 1,\ \cdots N-1\} \quad (9)$$

where θ_n is the nth orientation and N is the total number of orientations. In the study, a total of 40 features are extracted by selecting five scales and eight orientations. Then, to reduce the dimensionality of features and condense the relevant information, PCA is applied. The dimension of the features is compressed to six, a value which considers both the information of the features and the efficiency of computation [51]. Furthermore, as supplementary features, the mean and standard deviation, considering the information of neighboring pixels, are included. Here, to reflect both the coarse and fine-texture information of neighborhoods sufficiently, window sizes of 3×3, 5×5, 7×7, and 9×9 are selected.

2.2.4. Learning-Based Image Fusion

As mentioned above, there are significant differences in imaging mechanisms between SAR and panchromatic images. To consider the differences in the imaging mechanisms, compositive characteristics should be utilized, and non-linear relationships are required. Therefore, this study employed RF regression, which is a representative algorithm that considers compositive characteristics and models non-linear relationships. RF regression is based on the classification and regression tree (CART) model, which is an ensemble-based algorithm that combines several decision trees and obtains results [52]. For the classification, the results are obtained by most votes from the tree results, whereas for regression, the tree results are averaged [53]. In particular, each tree is created independently through a process called bootstrap aggregation, or bagging, to avoid correlations with other trees; in this process, training data subsets are drawn by randomly resampling the subsets with replacement from the original training dataset [54,55]. Thus, this process is robust to the presence of noise or slight variations in the input data, has greater stability, and increases the prediction accuracy [56,57]. Furthermore, in each tree, approximately 30% of the data is excluded from the training process, which is called out-of-bag (OOB) data. The mean squared error (MSE) between the OOB data and the data used for growing the regression trees is obtained; then, a prediction error called the OOB error is calculated for each variable [53]. This error estimates the importance of every variable, such that the influence of each input variable can be further analyzed. To determine the importance of the input variables, each variable is permuted, and regression trees are grown on the modified dataset [58]. The variable importance is calculated based on the difference in the MSE between the original OOB dataset and the modified dataset. In other words, if the exclusion of a variable leads to a significant reduction in prediction accuracy, the variable is considered important.

In addition, the RF algorithm requires the specification of two parameters: the number of variables to be used for the best split at each node (m_{try}) and the number of trees in the forest (n_{tree}). In regression problems, the standard value of m_{try} is one-third of the total number of input variables; thus, in this study, m_{try} was selected as 5 [59]. Regarding n_{tree}, previous studies have shown that using a large value for n_{tree} provides better stability. However, recent studies have revealed that n_{tree} has no significant effect on performance; thus, in this study, n_{tree} was selected as 32 considering both the performance and training time [28,34,52].

Using the two parameters, the RF is modeled and generated independently for each class, which leads to a reduction in the complexity of the algorithm and allows more information to be retrieved. For each class, supervised learning is performed by setting the features extracted from the SAR image as independent variables and the pixel values of the panchromatic image as dependent variables for the positions corresponding to previously obtained invariant pixels. Then, the features of the SAR image corresponding to all the pixels of each class are extracted and utilized as input values of the obtained RF regression model. Finally, the fusion result is generated by integrating the predicted values for each class.

2.3. Criteria for Fusion Quality Assessment

The quality of the image-fusion results can be evaluated according to two criteria. First, the performance of the fusion results for the proposed method can be intuitively evaluated in terms of visual aspect. Second, quantitative evaluation can be used to obtain the performance of fusion results, which must be statistical and objective [22]. To assess the performance, the fusion quality index (FQI), average gradient (AG), and spatial frequency (SF) are selected. FQI is an index for evaluating the quality of a fused image for given input images; it is based on the combination of luminance distortion, contrast distortion, and loss of correlation of coefficient over local regions into a single measure [60]. Given the SAR image I_S, the panchromatic image I_P, and the fused image I_F, FQI is defined as Equation (10):

$$FQI = \sum_{w \in W} c(w)(\lambda(w)QI(I_S, I_F|w) + (1 - \lambda(w))QI(I_P, I_F|w)) \tag{10}$$

where $\lambda(w) = \frac{{\sigma_{I_S}}^2}{{\sigma_{I_S}}^2 + {\sigma_{I_P}}^2}$ is a weight computed over a window w, in which ${\sigma_{I_S}}^2$ and ${\sigma_{I_P}}^2$ are the variance of the SAR and panchromatic images, respectively; $c(w) = \frac{C(w)}{\sum_{w' \in W} C(w')}$ is a saliency computed over a window w, where $C(w) = \max({\sigma_{I_S}}^2, {\sigma_{I_P}}^2)$; $QI(I_S, I_F|w)$ and $QI(I_P, I_F|w)$ are the quality indexes of the fused image with regard to the SAR and panchromatic images, respectively; and w is set to 8×8 [60]. Given two images a and b, the QI is defined as Equation (11):

$$QI = \frac{4\sigma_{ab}\, \mu_a\, \mu_b}{({\mu_a}^2 + {\mu_b}^2)({\sigma_a}^2 + {\sigma_b}^2)} \tag{11}$$

where μ_a and μ_b are the means of the respective images; σ_a and σ_b are the standard deviations of the respective images; and σ_{ab} is the covariance of the two images. AG represents information on the edge details of an image, which is sensitive to the details of contrast and texture in the image [22]; it is defined as Equation (15):

$$AG = \frac{1}{MN}\sum_{i=1}^{M}\sum_{j=1}^{N}\sqrt{\frac{\Delta I_{Fx}{}^{2} + \Delta I_{Fy}{}^{2}}{2}} \tag{12}$$

where ΔI_{Fx} and ΔI_{Fy} are the differences in the x and y directions in the fused image, respectively. SF reflects the active degree of an image in the spatial domain [61] and is defined as Equations (16)–(18):

$$Row\ frequency = \sqrt{\frac{1}{MN}\sum_{i=1}^{M}\sum_{j=2}^{N}(I_F(i,j) - I_F(i,j-1))^2} \tag{13}$$

$$Column\ frequency = \sqrt{\frac{1}{MN}\sum_{i=2}^{M}\sum_{j=1}^{N}(I_F(i,j) - I_F(i-1,j))^2} \tag{14}$$

$$SF = \sqrt{Row\ frequency^2 + Column\ frequency^2} \tag{15}$$

FQI has the range of $[0,\ 1]$, and an FQI value closer to 1 indicates better performance, whereas for AG and SF, higher values indicate better performance [61].

3. Results and Discussion

3.1. Comparison of Fusion Results

To evaluate the effectiveness of the proposed fusion approach, its results were compared with those of the conventional image-fusion algorithm. To ensure a fair comparison, fusion algorithms using a single SAR and panchromatic image were considered, where the à-trous wavelet decomposition (ATWD) [20], DWT [23], NSCT [19], and NSCT-pulse couple neural network (NSCT-PCNN) [5] methods were selected. The ATWD method is based on the importance of the wavelet coefficient, which is incorporated into the SAR image at a certain high frequency. For the DWT method, the maximum values of the coefficients at low frequencies and high frequencies are selected as the low and high frequencies, respectively. The NSCT method is based on the contourlet transform without downsamplers and upsamplers, and it also selects the averaging scheme at a low frequency and the maximum scheme at high frequency. The NSCT-PCNN method performs fusion based on PCNN for coefficients at low frequencies, and the coefficients at high frequencies are obtained through NSCT. In accordance with the aforementioned details, the decomposition level of NSCT and NSCT-PCNN was selected as the three showing the best image-fusion results [19]. Furthermore, the experiments were carried out on a desktop PC with an Intel(R) Core (TM) i7-8700 @ 3.20 GHz processor, 24.00 GB of RAM, and a 64-bit Windows 10 operating system. Particularly, all experiments involving the proposed model were programmed in Python 3.7, and the conventional methods were programmed in MATLAB 2019a. The image-fusion results are shown in Figures 6–8.

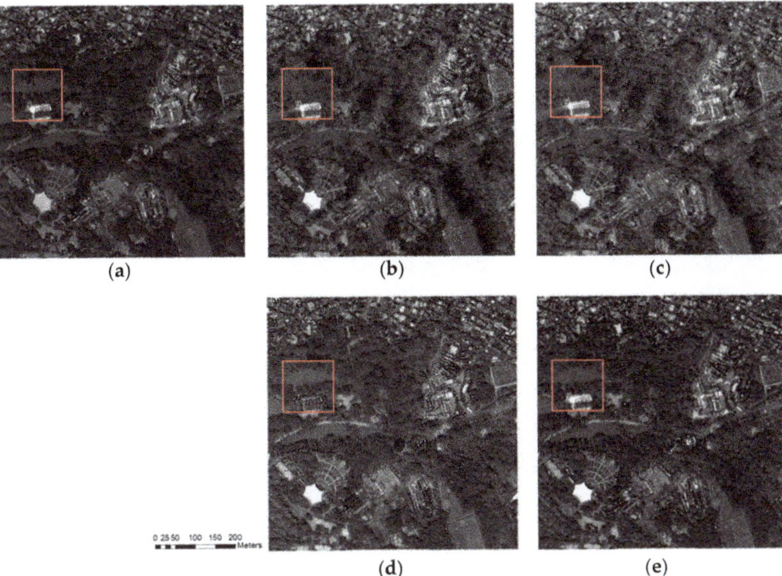

Figure 6. Comparison with the results of image fusion for Site 1: (**a**) proposed method; (**b**) a-trous wavelet decomposition; (**c**) discrete wavelet transform; (**d**) non-subsampled contourlet transform; and (**e**) non-subsampled contourlet transform-pulse couple neural network.

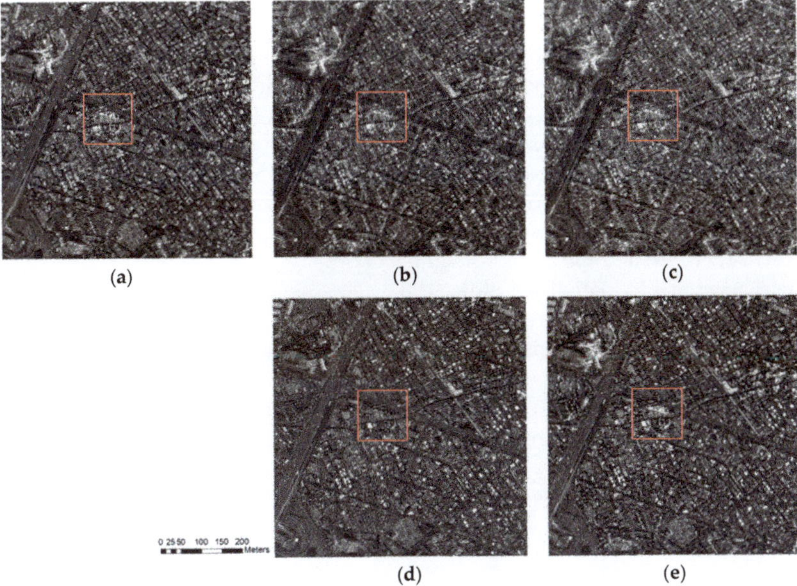

Figure 7. Comparison with the results of image fusion for Site 2: (**a**) proposed method; (**b**) a-trous wavelet decomposition; (**c**) discrete wavelet transform; (**d**) non-subsampled contourlet transform; and (**e**) non-subsampled contourlet transform-pulse couple neural network.

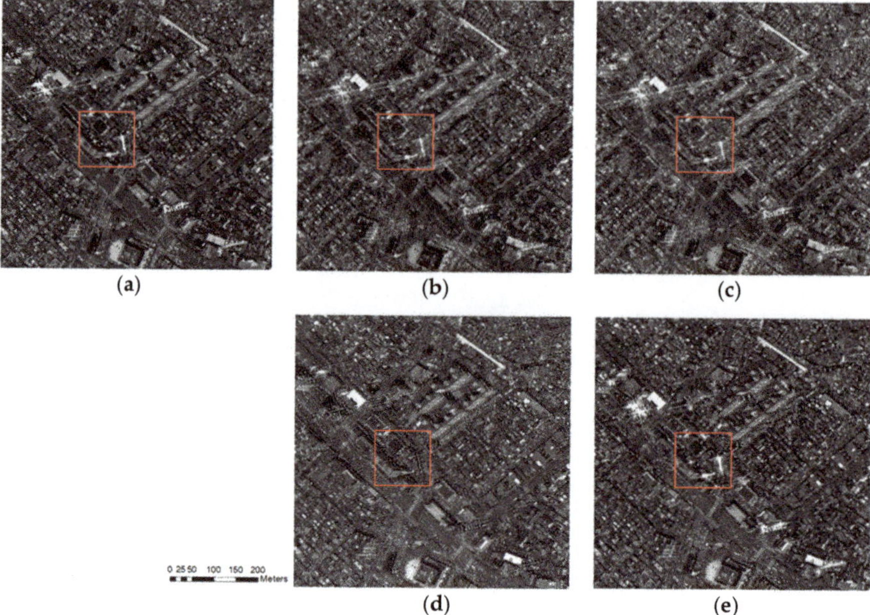

Figure 8. Comparison with the results of image fusion for Site 3: (**a**) proposed method; (**b**) a-trous wavelet decomposition; (**c**) discrete wavelet transform; (**d**) non-subsampled contourlet transform; and (**e**) non-subsampled contourlet transform-pulse couple neural network.

From an overall visual inspection, the results of the proposed method and those of the conventional fusion methods provided more information than the original single image. They contained spatial information of the panchromatic image, such as the line information and edge information of buildings, as well as the object information of the SAR image. However, in the results of ATWD and DWT, spatial information was insufficient compared with those of other methods. For Site 1, which primarily consisted of vegetation and included developed structures, the surface roughness of the SAR image in both these areas was emphasized, resulting in less spatial information. Sites 2 and 3 mainly consisted of developed structures, and the surface roughness of the SAR image was also emphasized more than the line and edge information of buildings, like the results for Site 1. Furthermore, more spatial information was present in the result of the NSCT than in the result of ATWD or DWT; however, it was also confirmed that the object information of the SAR image was lost compared to the original SAR image. The result of the NSCT-PCNN included more spatial and object information compared to those of the conventional image-fusion methods; however, the spatial information of the vegetation in Site 1 was somewhat insufficient. In contrast, the proposed method included sufficient spatial and object information regardless of vegetation or developed areas. The specific details are indicated on the red rectangle in Figures 6–8, and the enlarged areas are shown in Figure 9.

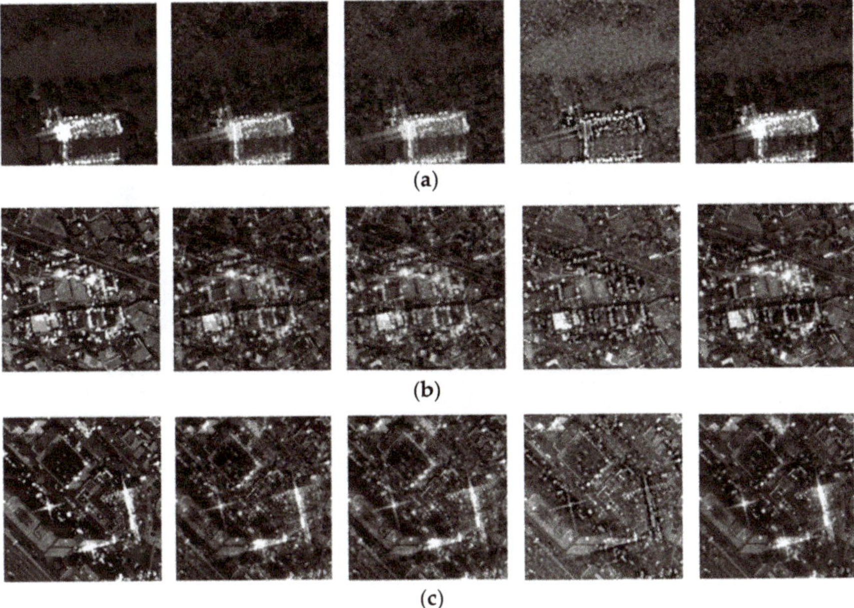

Figure 9. Enlargement of the area marked with a red rectangle: (**a**) Site 1, (**b**) Site 2, (**c**) Site 3. From left to right: proposed method, a-trous wavelet decomposition, discrete wavelet transforms, non-subsampled contourlet transform, and non-subsampled contourlet transform-pulse couple neural network.

Although visual analysis is direct and intuitive, it is also highly subjective and, therefore, may not allow for fully accurate evaluation. Thus, the performance of the fusion results was further evaluated quantitatively based on FQI, AG, and SF, which are summarized in Table 2. Regarding the FQI, the proposed method performed better than the conventional image-fusion methods in all sites. For site 1, the proposed method showed improvements of 8.51%, 6.24%, 27.14%, and 19.90% over ATWD, DWT, NSCT, and NSCT-PCAA, respectively, in addition to respective improvements of 2.78%, 0.53%, 21.92%, and 24.63% for Site 2 and respective improvements of 2.77%, 0.45%, 20.25%, and 14.68% for Site 3. The higher FQI of the proposed method indicates that its fusion results contain more of the information of the SAR and panchromatic images. In contrast, AG yielded different results for each site, as follows. For Site 1, the NSCT-PCNN had the highest value, whereas at Sites 2 and 3 the proposed method had the highest value. AG represents the spatial information in the panchromatic image in addition to the object information and surface roughness in the SAR image. As mentioned above, Site 1 consisted mostly of vegetation, and the result of the NSCT-PCNN contained most of the surface roughness information of the SAR image with a lack of the spatial information of the panchromatic image of the vegetation area. Because of this, the texture features of the vegetation were best highlighted owing to the influence of surface roughness in the calculation of AG. However, for Sites 2 and 3, which consisted mainly of developed structures, the spatial information of the panchromatic image and the object information of the SAR image were the main information, and the result of the proposed method had the highest abundance with regard to both aforementioned sets of information. Regarding the SF, which is primarily a metric for assessing the spatial information derived from the panchromatic image, the proposed method exhibited the best performance in all sites. In other words, it is confirmed that the proposed method would be more useful than the conventional image-fusion methods in visual and quantitative evaluations.

Table 2. Evaluations for the image fusion methods (FQI: fusion quality index, AG: average gradient, SF: spatial frequency, ATWD: à-trous wavelet decomposition, DWT: discrete wavelet transforms, NSCT: non-subsampled contourlet transform, NSCT-PCNN: non-subsampled contourlet transform-pulse couple neural network).

Site	Method	FQI	AG	SF
Site 1	Proposed method	0.8489	18.5836	10.1753
	ATWD	0.7638	14.8698	7.5353
	DWT	0.7865	14.3675	7.0175
	NSCT	0.5775	20.0136	8.4952
	NSCT-PCNN	0.6499	20.4469	9.2749
Site 2	Proposed method	0.8199	33.5763	17.0272
	ATWD	0.7921	21.1922	11.0085
	DWT	0.8146	21.1521	10.4409
	NSCT	0.6007	28.738	12.3708
	NSCT-PCNN	0.6527	30.0487	13.6092
Site 3	Proposed method	0.7936	29.9653	15.7126
	ATWD	0.7659	19.2859	10.3327
	DWT	0.7891	20.2947	10.2658
	NSCT	0.5911	25.9567	11.4126
	NSCT-PCNN	0.6468	27.9568	12.9668

3.2. Validation of Random Forest Regression

As mentioned above, the RF regression models were constructed independently for each class, thus, the predictive models were verified separately. The classification images of each site are shown in Figure 10, and the characteristics of each class are as follows: Classes 1 and 2 represent areas with high backscattered intensity and double bounce scattering characteristics, because of the artifacts in the SAR image, where the intensity of class 2 is lower than that of class 1. Classes 3 and 4 include the specular reflection characteristics of the SAR image and bare land or those of the high-brightness roofs in the panchromatic image, where class 4 is brighter than class 3 in the panchromatic image. Classes 5 and 6 are composed of vegetation, roads, or low buildings in the panchromatic image; the low backscattered intensity characteristics of class 5 and diffuse scattering characteristics of class 6 are shown in the SAR image.

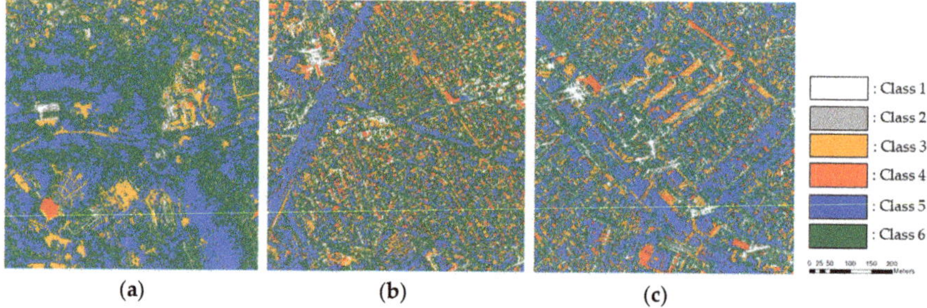

Figure 10. Classification images: (**a**) Site 1, (**b**) Site 2, (**c**) Site 3.

The evaluation was performed visually with a scatter plot and quantitatively using the coefficient of determination (R^2), as shown in Figure 11 and Table 3. The scatter plot represents the correlation between data, and there are high correlations among all classes regardless of the site. In particular, Site 1 showed a high correlation among classes 1, 2, and 4, whereas Sites 2 and 3 showed a high correlation between classes 1 and 4. The other classes showed a moderate bias but a sufficiently high correlation. Considering R^2, a high value of which indicates the high precision and accuracy of the model, similar tendencies are observed, as follows. For all sites, classes 1 and 4 had the highest R^2, and both exhibited similar properties. However, for class 2, Sites 2 and 3 were somewhat lower than Site 1, which is thought to be because of the complex structure of many buildings. Furthermore, classes 3, 5, and 6 involved several characteristics, which can lead to relatively low R^2 values. However, the overall results are reasonable; thus, the robustness of the constructed modes is confirmed.

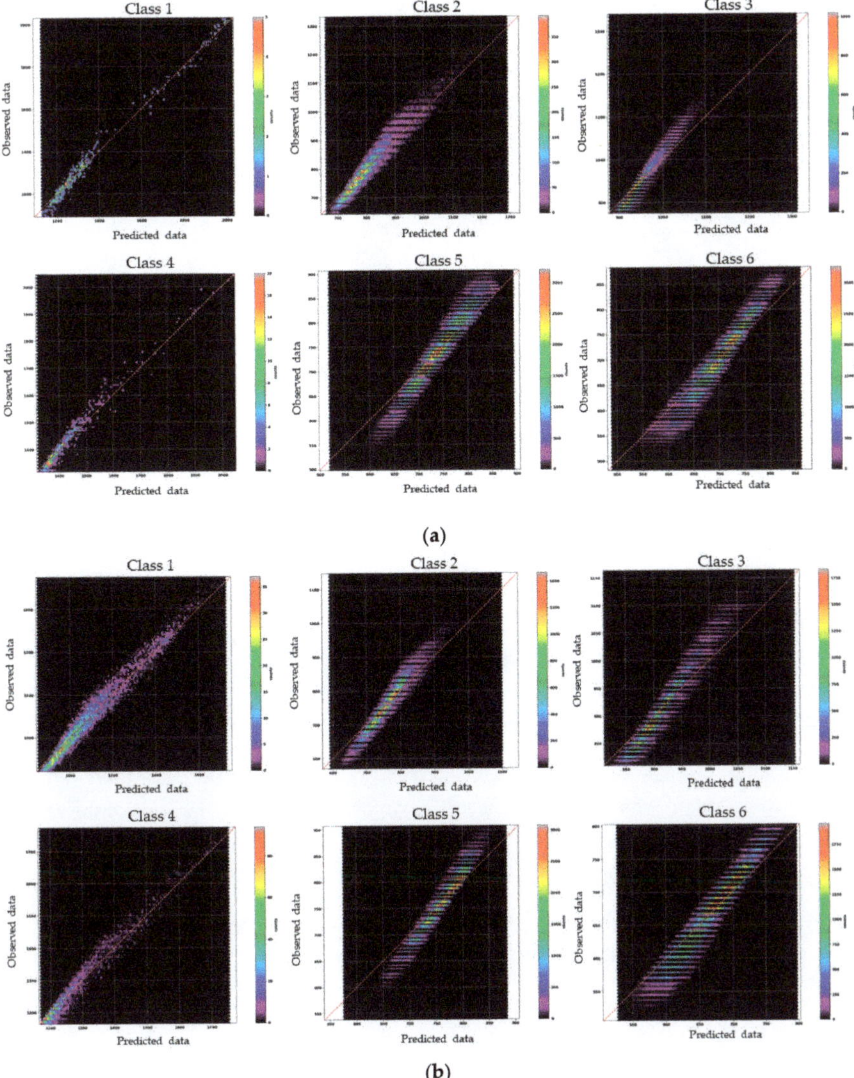

Figure 11. *Cont.*

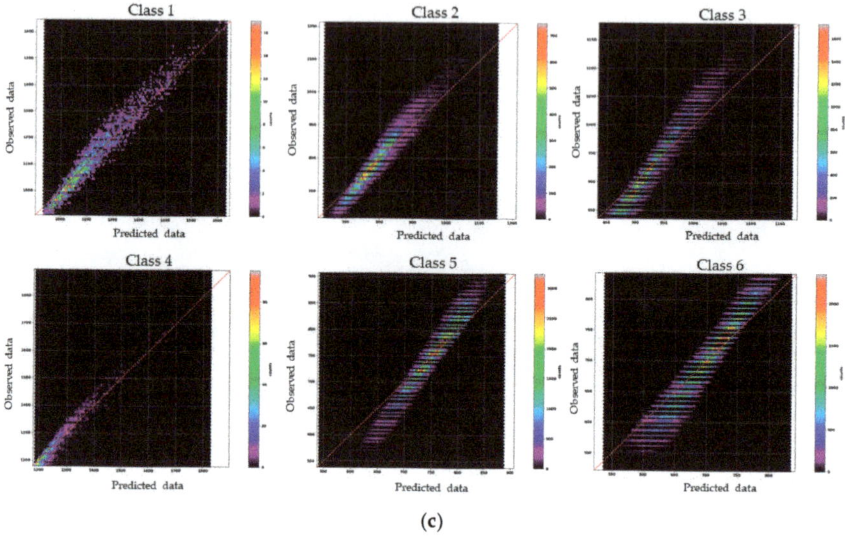

(**c**)

Figure 11. Scatter plots for each class: (**a**) Site 1, (**b**) Site 2 (**c**) Site 3.

Table 3. R^2 Values of the predictive models for each class.

Site	Class 1	Class 2	Class 3	Class 4	Class 5	Class 6
Site 1	0.9856	0.9209	0.9171	0.9745	0.8698	0.8771
Site 2	0.9594	0.8775	0.8791	0.9606	0.8584	0.8792
Site 3	0.9450	0.8917	0.8692	0.9591	0.8593	0.8717

3.3. Variable Importance

To evaluate the influence of variables on RF regression, the variable importance scores were obtained. In particular, the variable importance scores were evaluated for each site, and the importance of each variable was averaged by all classes, as shown in Table 4. In terms of the importance of an individual variable, regardless of the site, the intensity of the SAR image contributed the most, followed by the mean of window sizes 3×3, 5×5, 7×7, and 9×9. In the case of the Gabor filter and standard deviation, the contribution was approximately 4–6%, which was relatively insignificant. On the other hand, in terms of the variable type, the contributions of intensity, Gabor filter, mean, and standard deviation were approximately 13–16%, 25–30%, 36–37%, and 17–20%, respectively; thus, it is confirmed that all variables are properly influenced.

However, it should be noted that the variable importance scores are relative; therefore, they depend on the number of variables included. In other words, the importance scores can be changed by removing or replacing the predictors, as different inter-correlated variables could act as substitutes.

Table 4. Variable importance scores averaged across all classes.

Variable Importance Scores (%)	Site 1	Site 2	Site 3
Intensity	13.24	16.59	15.03
Gabor filter-principal component 1	5.70	5.72	5.94
Gabor filter-principal component 2	4.90	4.61	4.79
Gabor filter-principal component 3	4.88	4.60	4.80
Gabor filter-principal component 4	4.91	4.57	4.80
Gabor filter-principal component 5	4.92	4.55	4.78
Gabor filter-principal component 6	4.94	4.67	4.78
Mean (3 × 3)	12.82	12.27	13.45
Mean (5 × 5)	10.19	8.97	9.50
Mean (7 × 7)	6.92	8.49	7.82
Mean (9 × 9)	6.66	7.07	6.13
Standard deviation (3 × 3)	4.83	4.44	4.50
Standard deviation (5 × 5)	4.85	4.29	4.43
Standard deviation (7 × 7)	4.95	4.45	4.49
Standard deviation (9 × 9)	5.31	4.72	4.78

3.4. Additional Dataset

One additional dataset was included to verify the robustness of the proposed method. The area is St. John's, Newfoundland and Labrador, which is located along the Atlantic Ocean and mainly contains the water, grass, barren land, forest, and developed structures, and the panchromatic and SAR images are acquired from the GeoEye-1 and TerraSAR-X sensors. The GeoEye-1 image was acquired on 19 August 2019; it has a 0.46 m spatial resolution and 11-bit radiometric resolution. The TerraSAR-X was acquired on 8 August 2019; it was obtained in Staring SpotLight mode with a 0.8 m × 0.25 m spatial resolution, an ascending orbit, and HH polarization. The preprocessing was performed in the same way as that for the previously used dataset, elucidated in the previous sections, and the additional experiments were performed on two sites with a subset of 1500 × 1500 pixels. The additional experimental images and results are shown in Figures 12 and 13. From a visual inspection, it can be seen that the fusion was properly performed and that both the spatial information of the panchromatic image and the object information of the SAR image are sufficiently present in the resultant image. Furthermore, as shown in Table 5, the performance for the additional sites was like that in the previous results. That is, it is confirmed that the proposed method shows satisfactory results for the additional dataset, and its applicability is verified.

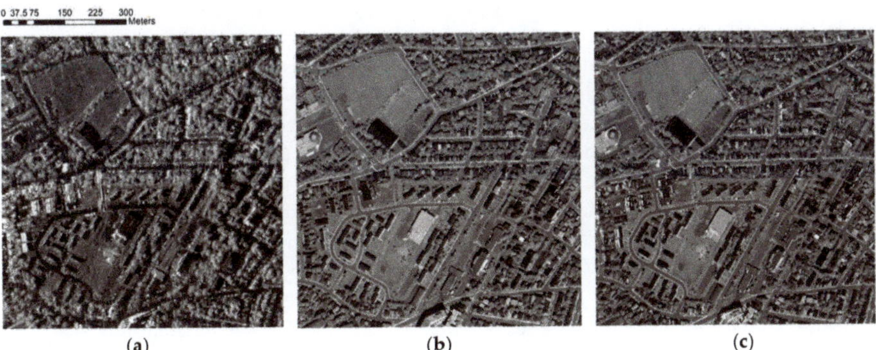

Figure 12. Experimental images in the additional dataset (Site 1): (**a**) TerraSAR-X image acquired on 8 August 2019, (**b**) GeoEye-1 image acquired on 19 August 2019, (**c**) fusion result of the proposed method.

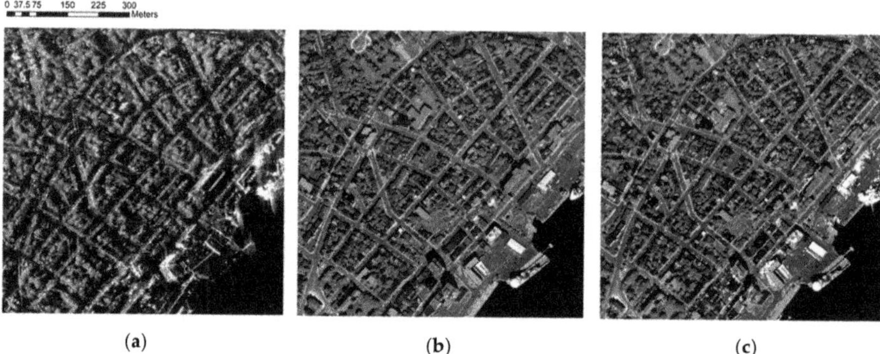

(a) (b) (c)

Figure 13. The experimental images in the additional dataset (Site 2): (**a**) TerraSAR-X image acquired on 8 August 2019, (**b**) GeoEye-1 image acquired on 19 August 2019, (**c**) fusion result of the proposed method.

Table 5. Evaluations of the additional dataset (FQI: fusion quality index, AG: average gradient, SF: spatial frequency).

Site	FQI	AG	SF
Site 1	0.7967	24.0456	14.8655
Site 2	0.7873	28.3798	18.0001

4. Conclusions

This study proposes a method that fuses high-resolution SAR and panchromatic images. A learning-based approach is adopted, and RF regression, which considers the differences in imaging mechanisms, forms the basis of the proposed method. The proposed method first selects the pixels to be used for learning and then performs classification on stacked SAR and panchromatic images to establish independent relationships for each class, thereby reducing the algorithm complexity. In particular, the number of classes is selected as six considering the land cover distributions and training time. Furthermore, to consider as many features as possible, various features are extracted from the SAR image, among which the Gabor filter and the mean and standard deviation of multiple window sizes are selected. Finally, image fusion is performed based on RF regression; then, the results are compared with those of conventional image-fusion methods. The following conclusions are obtained based on the results. First, from the visual aspect, the proposed method includes more of the object information of the SAR image and spatial information of the panchromatic image than conventional image-fusion methods. It is confirmed that sufficient information is included, regardless of vegetation and built-up areas. Second, the quantitative performance of the proposed method shows significant improvements. The performance evaluation verifies that the proposed method exhibits improved preservation of the information of the SAR and panchromatic images and results in less distortion when compared with conventional image-fusion methods. Third, when validating the RF regression model employed in the proposed method, it is confirmed that the predictive model is properly constructed. In addition, in the case of the variables selected, they contribute appropriately to the RF regression model. Finally, the applicability of the proposed model is verified by applying the proposed method to an additional dataset.

In future studies, the following aspects should be considered. First, by obtaining and applying the method to a sufficiently wide range of seasonal and temporal images, it should be further verified. Second, the method's usefulness should be further confirmed through application to SAR and panchromatic images obtained from other sensors. Third, the performance of the RF regression process should be improved by further extracting and combining various features. Finally, its applicability should be investigated by applying the fused images to various applications.

Author Contributions: Conceptualization, Y.D.E.; methodology, D.K.S.; software, D.K.S.; validation, D.K.S; formal analysis, D.K.S.; investigation, D.K.S.; resources, D.K.S.; data curation, Y.D.E.; writing—original draft preparation, Y.D.E.; writing—review and editing, Y.D.E.; visualization, D.K.S.; supervision, Y.D.E.; project administration, Y.D.E.; funding acquisition, Y.D.E. All authors have read and agreed to the published version of the manuscript.

Funding: This work was supported by the National Research Foundation of Korea (NRF) grant funded by the Korea government (MSIT) (No. 2019R1A2C1085618).

Conflicts of Interest: The authors declare no conflict of interest.

References

1. Chibani, Y. Integration of panchromatic and SAR features into multispectral SPOT images using the 'à trous' wavelet decomposition. *Int. J. Remote Sens.* **2007**, *28*, 2295–2307. [CrossRef]
2. Amarsaikhan, D.; Blotevogel, H.; Van Genderen, J.; Ganzorig, M.; Gantuya, R.; Nergui, B. Fusing high-resolution SAR and optical imagery for improved urban land cover study and classification. *Int. J. Image Data Fusion* **2010**, *1*, 83–97. [CrossRef]
3. Hong, G.; Zhang, Y.; Mercer, B. A Wavelet and HIS Integration Method to Fuse High Resolution SAR with Moderate Resolution Multispectral Images. *Photogramm. Eng. Remote Sens.* **2009**, *75*, 1213–1223. [CrossRef]
4. Krylov, V.A.; Moser, G.; Voisin, A.; Serpico, S.B.; Zerubia, J. Change detection with synthetic aperture radar images by Wilcoxon statistic likelihood ratio test. In Proceedings of the 2012 19th IEEE International Conference on Image Processing, Orlando, FL, USA, 30 September–3 October 2012; pp. 2093–2096.
5. Mercier, G.; Moser, G.; Serpico, S. Conditional Copulas for Change Detection in Heterogeneous Remote Sensing Images. *IEEE Trans. Geosci. Remote Sens.* **2008**, *46*, 1428–1441. [CrossRef]
6. Wang, X.L.; Chen, C.X. Image fusion for synthetic aperture radar and multispectral images based on sub-band-modulated non-subsampled contourlet transform and pulse coupled neural network methods. *Imaging Sci. J.* **2016**, *64*, 87–93. [CrossRef]
7. Pappas, O.; Achim, A.; Bull, D.R. Curvelet fusion of panchromatic and SAR satellite imagery using fractional lower order moments. In Proceedings of the 2013 10th IEEE International Conference on Advanced Video and Signal Based Surveillance, Krakow, Poland, 27–30 August 2013; pp. 342–346.
8. Gungor, O.; Shan, J. An Optimal Fusion Approach for Optical and SAR Images. In Proceedings of the ISPRS Commission VII Symposium: Remote Sensing: From Pixels to Process, Enschede, The Netherlands, 8–11 May 2006; pp. 111–116.
9. Reyes, M.F.; Auer, S.J.; Merkle, N.; Henry, C.; Schmitt, M. SAR-to-Optical Image Translation Based on Conditional Generative Adversarial Networks—Optimization, Opportunities and Limits. *Remote Sens.* **2019**, *11*, 2067. [CrossRef]
10. Li, Y.; Fu, R.; Meng, X.; Jin, W.; Shao, F. A SAR-to-Optical Image Translation Method Based on Conditional Generation Adversarial Network (cGAN). *IEEE Access* **2020**, *8*, 60338–60343. [CrossRef]
11. Hellwich, O.; Heipke, C.; Wessel, B. Sensor and data fusion contest: Information for mapping from airborne SAR and optical imagery. In Proceedings of the IGARSS 2001, Scanning the Present and Resolving the Future, Proceedings, IEEE 2001 International Geoscience and Remote Sensing Symposium (Cat. No.01CH37217), Sydney, Australia, 9–13 July 2001; Volume 6, pp. 2793–2795.
12. Santoso, A.W.; Bayuaji, L.; Sze, L.T.; Lateh, H.; Zain, J.M. Comparison of Various Speckle Noise Reduction Filters on Synthetic Aperture Radar Image. *Int. J. Appl. Eng. Res.* **2016**, *11*, 8760–8767.
13. Zeng, Y.; Zhang, J.; Van Genderen, J.L. Change Detection Approach to SAR and Optical Image Integration. *Int. Arch. Photogramm. Remote Sens. Spat. Inf. Sci.* **2008**, *XXXVII*, 1077–1084.
14. Shimada, M. Ortho-Rectification and Slope Correction of SAR Data Using DEM and Its Accuracy Evaluation. *IEEE J. Sel. Top. Appl. Earth Obs. Remote Sens.* **2010**, *3*, 657–671. [CrossRef]
15. Al-Nasrawi, A.K.M.; Hopley, C.A.; Hamylton, S.; Jones, B.G. A Spatio-Temporal Assessment of Landcover and Coastal Changes at Wandandian Delta System, Southeastern Australia. *J. Mar. Sci. Eng.* **2017**, *5*, 55. [CrossRef]
16. Klonus, D. Comparison of Pansharpening Algorithms for Combining Radar and Multispectral Data. *Int. Arch. Photogramm. Remote Sens. Spat. Inf. Sci.* **2008**, *XXXVII*, 189–194.
17. Orsomando, F.; Lombardo, P.; Zavagli, M.; Costantini, M. SAR and Optical Data Fusion for Change Detection. In Proceedings of the Urban Remote Sensing Joint Event, Paris, France, 11–17 April 2007; pp. 1–9.

18. Abdikan, S.; Sanli, F.B. Comparison of different fusion algorithms in urban and agricultural areas using sar (palsar and radarsat) and optical (spot) images. *Boletim de Ciências Geodésicas* **2012**, *18*, 509–531. [CrossRef]
19. Ye, C.; Zhang, L.; Zhang, Z. SAR and panchromatic image fusion based on region features in nonsubsampled contourlet transform domain. In Proceedings of the 2012 IEEE International Conference on Automation and Logistics, Zhengzhou, China, 15–17 August 2012; Institute of Electrical and Electronics Engineers (IEEE): Piscataway, NJ, USA, 2012; pp. 358–362.
20. Pohl, C.; Van Genderen, J. Review article Multisensor image fusion in remote sensing: Concepts, methods and applications. *Int. J. Remote Sens.* **1998**, *19*, 823–854. [CrossRef]
21. Chibani, Y. Selective Synthetic Aperture Radar and Panchromatic Image Fusion by Using the à Trous Wavelet Decomposition. *EURASIP J. Adv. Signal Process* **2005**, *2005*, 404562. [CrossRef]
22. Liu, G.; Li, L.; Gong, H.; Jin, Q.; Li, X.; Song, R.; Chen, Y.; Chen, Y.; He, C.; Huang, Y.; et al. Multisource Remote Sensing Imagery Fusion Scheme Based on Bidimensional Empirical Mode Decomposition (BEMD) and Its Application to the Extraction of Bamboo Forest. *Remote Sens.* **2016**, *9*, 19. [CrossRef]
23. Pajares, G.; De La Cruz, J.M. A wavelet-based image fusion tutorial. *Pattern Recognit.* **2004**, *37*, 1855–1872. [CrossRef]
24. Ma, X.; Hu, S.; Liu, S.; Fang, J.; Xu, S. Remote Sensing Image Fusion Based on Sparse Representation and Guided Filtering. *Electronics* **2019**, *8*, 303. [CrossRef]
25. Miao, Q.; Wang, B. A Novel Fusion Method Using Contourlet Transform. In Proceedings of the 2006 International Conference on Communications, Circuits and Systems, Guilin, China, 25–28 June 2006; pp. 548–552.
26. Da Cunha, A.L.; Zhou, J.; Do, M.N. The nonsubsampled contourlet transform: Theory, design, and applications. *IEEE Trans. Image Process* **2006**, *15*, 3089–3101. [CrossRef]
27. Zhu, Z.; Yin, H.; Chai, Y.; Li, Y.; Qi, G. A novel multi-modality image fusion method based on image decomposition and sparse representation. *Inf. Sci.* **2018**, *432*, 516–529. [CrossRef]
28. Seo, D.; Kim, Y.; Eo, Y.D.; Lee, M.H.; Park, W.Y. Fusion of SAR and Multispectral Images Using Random Forest Regression for Change Detection. *ISPRS Int. J. Geo Inf.* **2018**, *7*, 401. [CrossRef]
29. Hultquist, C.; Chen, G.; Zhao, K. A comparison of Gaussian process regression, random forests and support vector regression for burn severity assessment in diseased forests. *Remote Sens. Lett.* **2014**, *5*, 723–732. [CrossRef]
30. Tsai, F.; Lai, J.-S.; Lu, Y.-H. Full-Waveform LiDAR Point Cloud Land Cover Classification with Volumetric Texture Measures. *Terr. Atmos. Ocean. Sci.* **2016**, *27*, 549. [CrossRef]
31. Wang, B.; Choi, J.; Choi, S.; Lee, S.; Wu, P.; Gao, Y. Image Fusion-Based Land Cover Change Detection Using Multi-Temporal High-Resolution Satellite Images. *Remote Sens.* **2017**, *9*, 804. [CrossRef]
32. Hong, S.H.; Lee, K.Y.; Kim, Y.S. KOMPSAT-5 SAR Application. In Proceedings of the 2011 3rd International Asia-Pacific Conference on Synthetic Aperture Radar (APSAR), Seoul, Korea, 26–30 September 2011; pp. 1–2.
33. Dimov, D.; Kuhn, J.; Conrad, C. Assessment of Cropping System Diversity in the Fergana Valley through Image Fusion of Landsat 8 and Sentinel-1. In Proceedings of the XXIII ISPRS Congress, Prague, Czech Republic, 12–19 July 2016; pp. 173–180.
34. Seo, D.; Kim, Y.; Eo, Y.D.; Park, W.Y. Learning-Based Colorization of Grayscale Aerial Images Using Random Forest Regression. *Appl. Sci.* **2018**, *8*, 1269. [CrossRef]
35. Ghosh, A.; Mishra, N.S.; Ghosh, S. Fuzzy clustering algorithms for unsupervised change detection in remote sensing images. *Inf. Sci.* **2011**, *181*, 699–715. [CrossRef]
36. Shao, P.; Shi, W.; He, P.; Hao, M.; Zhang, X. Novel Approach to Unsupervised Change Detection Based on a Robust Semi-Supervised FCM Clustering Algorithm. *Remote Sens.* **2016**, *8*, 264. [CrossRef]
37. Yan, W.; Shi, S.; Pan, L.; Zhang, G.; Wang, L. Unsupervised change detection in SAR images based on frequency difference and a modified fuzzy c-means clustering. *Int. J. Remote Sens.* **2018**, *39*, 3055–3075. [CrossRef]
38. Seo, D.K.; Eo, Y.D. Relative Radiometric Normalization for High-Resolution Satellite Imagery Based on Multilayer Perceptron. *J. Korean Soc. Surv. Geod. Photogramm Cartogr.* **2018**, *36*, 515–523. [CrossRef]
39. Seo, D.; Eo, Y.D. Multilayer Perceptron-Based Phenological and Radiometric Normalization for High-Resolution Satellite Imagery. *Appl. Sci.* **2019**, *9*, 4543. [CrossRef]
40. Zakeri, H.; Yamazaki, F.; Liu, W. Texture Analysis and Land Cover Classification of Tehran Using Polarimetric Synthetic Aperture Radar Imagery. *Appl. Sci.* **2017**, *7*, 452. [CrossRef]

41. Vigneshl, T.; Thyagharajan, K.K. Local binary pattern texture feature for satellite imagery classification. In Proceedings of the 2014 International Conference on Science Engineering and Management Research (ICSEMR), Chennia, India, 27–29 November 2014; pp. 1–6.
42. Huang, C. Terrain classification of polarimetric synthetic aperture radar imagery based on polarimetric features and ensemble learning. *J. Appl. Remote Sens.* **2017**, *11*, 26002. [CrossRef]
43. Marmel, U. Use of Filters for Texture Classification of Airborne Images and LIDAR data. *Arch. Photogramm. Cartogr. Remote Sens.* **2011**, *22*, 325–336.
44. Kim, J.; Um, S.; Min, D. Fast 2D Complex Gabor Filter With Kernel Decomposition. *IEEE Trans. Image Process* **2018**, *27*, 1713–1722. [CrossRef]
45. Chen, L.; Zhu, Q.; Xie, X.; Hu, H.; Zeng, H. Road Extraction from VHR Remote-Sensing Imagery via Object Segmentation Constrained by Gabor Features. *ISPRS Int. J. Geo Inf.* **2018**, *7*, 362. [CrossRef]
46. A Clausi, D.; Jernigan, M.E. Designing Gabor filters for optimal texture separability. *Pattern Recognit.* **2000**, *33*, 1835–1849. [CrossRef]
47. Chen, C.; Li, W.; Su, H.; Liu, K. Spectral-Spatial Classification of Hyperspectral Image Based on Kernel Extreme Learning Machine. *Remote Sens.* **2014**, *6*, 5795–5814. [CrossRef]
48. Kamarainen, J.-K.; Kyrki, V.; Kälviäinen, H. Invariance properties of gabor filter-based features-overview and applications. *IEEE Trans. Image Process* **2006**, *15*, 1088–1099. [CrossRef]
49. Bianconi, F.; Fernández, A. Evaluation of the effects of Gabor filter parameters on texture classification. *Pattern Recognit.* **2007**, *40*, 3325–3335. [CrossRef]
50. Ilonen, J.; Kamarainen, J.-K.; Kalviainen, H. Fast extraction of multi-resolution Gabor features. In Proceedings of the 14th International Conference on Image Analysis and Processing (ICIAP 2007), Modena, Italy, 10–14 September 2007; Institute of Electrical and Electronics Engineers (IEEE): Piscataway, NJ, USA, 2007; pp. 481–486.
51. Deng, H.B.; Lian, W.J.; Zhen, L.X.; Huang, J.C. A New Facial Expression Recognition Method Based on Local Filter Bank and PCA plus LDA. *Int. J. Inf. Tech.* **2005**, *11*, 86–96.
52. Seo, D.; Kim, Y.; Eo, Y.D.; Park, W.Y.; Park, H.C. Generation of Radiometric, Phenological Normalized Image Based on Random Forest Regression for Change Detection. *Remote Sens.* **2017**, *9*, 1163. [CrossRef]
53. Breiman, L. Random Forests. *Mach. Learn.* **2001**, *45*, 5–32. [CrossRef]
54. Peters, J.; De Baets, B.; Verhoest, N.E.C.; Samson, R.; Degroeve, S.; De Becker, P.; Huybrechts, W. Random forests as a tool for ecohydrological distribution modelling. *Ecol. Model.* **2007**, *207*, 304–318. [CrossRef]
55. Prasad, A.; Iverson, L.R.; Liaw, A. Newer Classification and Regression Tree Techniques: Bagging and Random Forests for Ecological Prediction. *Ecosystems* **2006**, *9*, 181–199. [CrossRef]
56. Rodriguez-Galiano, V.; Sanchez-Castillo, M.; Olmo, M.C.; Chica-Rivas, M. Machine learning predictive models for mineral prospectivity: An evaluation of neural networks, random forest, regression trees and support vector machines. *Ore Geol. Rev.* **2015**, *71*, 804–818. [CrossRef]
57. Hutengs, C.; Vohland, M. Downscaling land surface temperatures at regional scales with random forest regression. *Remote Sens. Environ.* **2016**, *178*, 127–141. [CrossRef]
58. Pal, M.; Singh, N.; Tiwari, N. Pier scour modelling using random forest regression. *ISH J. Hydraul. Eng.* **2013**, *19*, 69–75. [CrossRef]
59. Chagas, C.D.S.; Júnior, W.D.C.; Bhering, S.B.; Filho, B.C. Spatial prediction of soil surface texture in a semiarid region using random forest and multiple linear regressions. *Catena* **2016**, *139*, 232–240. [CrossRef]
60. Piella, G.; Hejimans, H. A New Quality for Image Fusion. In Proceedings of the 200 International Conference on Image Processing, Barcelona, Spain, 14–17 September 2003; pp. 173–176.
61. Pandit, V.R.; Bhiwani, R.J. Image Fusion in Remote Sensing Applications: A Review. *Int. J. Comput. Appl.* **2015**, *120*, 22–32. [CrossRef]

MDPI
St. Alban-Anlage 66
4052 Basel
Switzerland
Tel. +41 61 683 77 34
Fax +41 61 302 89 18
www.mdpi.com

Applied Sciences Editorial Office
E-mail: applsci@mdpi.com
www.mdpi.com/journal/applsci

www.ingramcontent.com/pod-product-compliance
Lightning Source LLC
LaVergne TN
LVHW070636170726
843515LV00004B/258

* 9 7 8 3 0 3 6 5 3 5 7 9 1 *